# Comprehensive Quality by Design in Pharmaceutical Development and Manufacture



Topical Conference at the 2009 AIChE Annual Meeting

## Nashville, Tennessee, USA
## 8-13 November 2009

ISBN: 978-1-61567-924-9

**Printed from e-media with permission by:**

Curran Associates, Inc.
57 Morehouse Lane
Red Hook, NY 12571

**Some format issues inherent in the e-media version may also appear in this print version.**



Printed by Curran Associates, Inc. (2010)

For permission requests, please contact AIChE
at the address below.

AIChE
3 Park Avenue
New York, NY 10016-5991

Phone: (203) 702-7660
Fax: (203) 775-5177

www.aiche.org

**Additional copies of this publication are available from:**

Curran Associates, Inc.
57 Morehouse Lane
Red Hook, NY 12571 USA
Phone: 845-758-0400
Fax: 845-758-2634
Email: curran@proceedings.com
Web: www.proceedings.com

# TABLE OF CONTENTS

# Progress in Implementation of Quality by Design - An FDA Perspective

Christine M.V. Moore, FDA/Cder/Ondqa, Food and Drug Administration, Silver Spring, MD

Quality by Design (QbD) is a science and risk based approach for pharmaceutical process development and manufacturing that begins with first defining clinically relevant product attributes and then designing and implementing a process to consistently deliver quality product. Some key aspects of QbD implementation are: (a) execution of a comprehensive risk assessment to identify material and/or process parameters that affect finished product quality, (b) understanding of multivariate interactions of process parameters and material attributes, and (c) implementation of a robust control strategy to ensure consistent quality. Knowledge management is pivotal for successful implementation of QbD to ensure that process understanding is routinely monitored and updated throughout the product life cycle.

Since the launch of the Office of New Drug Quality Assessment (ONDQA) CMC pilot program in July 2005, many QbD containing applications have been submitted, reviewed, and approved, both within and outside of the pilot. The objective of this presentation is to provide insight about the current state of QbD implementation as garnered from the QbD applications submitted to date. In addition, some remaining scientific and regulatory challenges will be discussed.

See more of this Session: QbD Topical Conference Plenary Session: Keynote Addresses
See more of this Group/Topical: Topical I: Comprehensive Quality by Design in Pharmaceutical Development and Manufacture

# The Evolution of Quality by Design in Drug Substance Development

**John Lepore**, Chemical Process Development and Commercialization, Merck & Co., Inc., Rahway, NJ

See more of this Session: QbD Topical Conference Plenary Session: Keynote Addresses
See more of this Group/Topical: Topical I: Comprehensive Quality by Design in Pharmaceutical Development and Manufacture

**TITLE**: **Drug Product Quality by Design: An Industrial Perspective**

**AUTHORS**: Mary T. am Ende[1]

**ABSTRACT**:

The regulatory quality guidances (Q8 – Q10) have set the expectation to develop pharmaceutical products based on a comprehensive quality by design approach. These expectations have permeated our industry as the new gold standard. Pharmaceutical development and commercialization programs are now well-versed in risk assessment processes used to prioritize and focus efforts in a science-based and rigorous manner to enable an intentionally designed quality product. The multivariate nature of drug products has posed a significant challenge to the industry, with respect to quality by design, that must now more aggressively leverage the knowledge of pharmaceutical scientists to rationalize product designs via science-based approach. The deployment of computational and predictive methods have progressed in recent years as the industry continues to drive development costs down by eliminating or reducing experimentation where appropriate. These efforts are augmenting the scientific understanding of drug product processes and scale-up. However, their use creates an additional challenge to establish the confidence of the prediction and how that translates to the design space. The purpose of this presentation is to discuss the future scientific gaps that must be addressed to develop a drug product using a comprehensive QbD approach, with reflections on the current state.

**KEYWORDS**: holistic design space, quality by design, DP process modeling, confidence in models and design space

ADDRESS OF AFFILIATION: [1]Pfizer Global Research & Development
Eastern Point Road, MS 8156-01
Groton, Connecticut 06340

TELEPHONE / FAX: (860) 441-4210
(860) 686-5404 (Fax)

email address: mary.t.am.ende@pfizer.com

# Analysis of Electrical Properties of a Powder Blend and Their Effect On Flow Behavior - a QbD Approach

Fernando J. Muzzio, Chemical & Biochemical Engineering, Rutgers University, Piscataway, NJ

See more of this Session: QbD Topical Conference Plenary Session: Keynote Addresses
See more of this Group/Topical: Topical I: Comprehensive Quality by Design in Pharmaceutical Development and Manufacture

# Modeling Batch Pharmaceutical Reactions by Factor Analysis of Spectral Data

**Fredy Afif**, Systems Research Institute and Chemical & Biological Engineering, Tufts University, Medford, MA

Christos Georgakis, Systems Research Institute and Chemical & Biological Engineering, Tufts University, Medford, MA

The availability of a variety of spectral measurement techniques (Raman, IR, NIR, et al) has enabled the collection of a plethora of data during the operation of batch reactions and many other processes. However, such spectral data is often influenced by several reaction components, the mixture temperature, and possibly the solvent. Consequently, a linear regressive model (e.g. via PLS), often called a Chemometric (CM) Model, is developed to relate spectral measurements to the respective mixture compositions. The development of such a model requires initial data for offline samples from the reacting mixture that includes spectral as well as composition information. The building of such a CM model is laborious and time consuming, especially when the temperature of the reactor is time-varying.

The present communication introduces a methodology that enables the modeling of the nonlinear time evolution of the reaction without the use of a CM model. Principal Component Analysis (PCA) of the spectral data at different times and from several batch runs reveals the time-invariant as well the time-variant characteristics of the reaction(s). Structured target factor analysis (Fotopoulos, Georgakis, & Stenger, 1994, 1998) of the time-invariant information helps define the possible reaction stoichiometry structure. The time-variant data is then used to develop the "kinetic expression" of the model using the generalized tendency kinetic modeling approach presented earlier (Makrydaki & Georgakis, 2007). This model is seen primarily as an aid for the optimization of the process, rather than the complete understanding of the chemical phenomena involved. The approach is tested against experimental data from a model pharmaceutical reaction.

---

Fotopoulos, J., Georgakis, C., & Stenger, H. G. (1994, June 28 - July 1). Structured target factor analysis for the stoichiometric modeling of batch reactors. Paper presented at the American Control Conference, Baltimore, MD.

Fotopoulos, J., Georgakis, C., & Stenger, H. G. (1998). Use of tendency models and their uncertainty in the design of state estimators for batch reactors. Chemical Engineering and Processing, 37(6), 545-558.

Makrydaki, F., & Georgakis, C. (2007). Multivariate Linear Regression as a Tool in Modeling the Kinetics of Complex Chemical Reactions, Annual AIChE meeting. Salt Lake City, UT.

See more of this Session: Qbd Across Drug Substance Operations
See more of this Group/Topical: Topical I: Comprehensive Quality by Design in Pharmaceutical Development and Manufacture

# Incorporating "Process Efficiency" Attributes and Lean Thinking During API Product Development"

**Ken Ryan**, Development Science and Technology, Pfizer Pharmaceuticals, Dublin 24, Ireland

Vincent Mc Curdy, Development Science and Technology, Pfizer Pharmaceuticals

Roger Weaver, Development Science and Technology, Pfizer Pharmaceuticals

Garry O'Connor, Development API, Pfizer Pharmaceuticals

Risk assessments are routinely used within Pfizer during product development to help teams identify areas of concern, prioritize experimentation and collate process understanding. In addition to focusing on QbD-type attributes, there is value in understanding process efficiency-type attributes (e.g. for API - filterability, phase separation, ease of cleaning, etc) and identifying those parameters and attributes which may have an impact. Examples will be presented which demonstrate the value of such an exercise. Lean manufacturing is concerned with reducing lead time, inventory and non value added work across a product value stream. Applying "lean" principles during product development (e.g. telescoping steps, employing PAT) has many benefits, examples of which will also be presented.

See more of this Session: Qbd Across Drug Substance Operations
See more of this Group/Topical: Topical I: Comprehensive Quality by Design in Pharmaceutical Development and Manufacture

# The Development of the Design Space for Maraviroc API

**Garry O'Connor**, Chemical Research & Development, Pfizer, Sandwich, United Kingdom

Ken Ryan, Development Science and Technology, Pfizer Pharmaceuticals, Dublin 24, Ireland

The presentation will describe how QbD principles were practically implemented during the transfer of the maraviroc API manufacturing process into commercial facilities. The presentation will demonstrate how the principles of QbD such as a risk based approach, design of experiments, design space development and six sigma tools were utilised to deliver an efficient, stable and approved manufacturing process exemplified by one of the most challenging steps in the process.

See more of this Session: Qbd Across Drug Substance Operations
See more of this Group/Topical: Topical I: Comprehensive Quality by Design in Pharmaceutical Development and Manufacture

# Linking Unit Operations During Design Space Development and Definition

Lynn Conley, Department of Process Biochemistry, Biogen Idec, San Diego, CA

Leonard Perry, University of San Diego, San Diego, CA

John Pieracci, Department of Process Biochemistry, Biogen Idec, San Diego, CA

**Matthew Westoby**, Department of Process Biochemistry, Biogen Idec, San Diego, CA

A greater importance has been placed on having a systemic and comprehensive approach to pharmaceutical development and lifecycle management as outlined by quality by design principles. Greater emphasis has been placed on defining the manufacturing design space. A drug substance process design space can be defined using a statistical based design of experiments (DOE) approach to establish input parameter ranges that provide assurance of consistent performance and meet product quality specifications. Traditionally, a separate set of designed experiments has been used for each unit operation to establish a robust operating space. Unfortunately, the performance of one step may depend on the quality attributes of the preceding process stream thus the linkage of process steps when conducting process characterization and defining the process design space is critical. In this presentation, we evaluate and compare two methods to link multiple unit operations during process characterization: forward linking and partition designs. Results of these methods applied to a biopharmaceutical process are presented and the benefits and challenges of each method are discussed.

See more of this Session: Qbd Across Drug Substance Operations
See more of this Group/Topical: Topical I: Comprehensive Quality by Design in Pharmaceutical Development and Manufacture

*AIChE 2009 Annual Meeting QbD Topical Conference Proceedings Paper (**111e**):*

# Quality-by-Design (QbD): Process Trajectory Development for a Pharmaceutical Co-precipitation Process Based on an Integrated Real-Time Process Monitoring Strategy

Huiquan Wu* and Mansoor A. Khan
Division of Product Quality Research (DPQR, HFD-940)
OTR/OPS/CDER/FDA
The FDA White Oak Campus, Life Science Building 64
10903 New Hampshire Ave, Silver Spring, MD 20993

Abstract: Pharmaceutical co-precipitation as an important technique to improve the dissolution and absorption of poorly soluble drugs has been modified to prepare extended release preparations. However, no much attention has been paid to demonstrate the process understanding of this important process. Process knowledge is an essential part of the Quality-by-Design (QbD) design space establishment. In this work, an integrated PAT approach was developed for real-time monitoring of a pharmaceutical (naproxen) and a polymer (eudragit) co-precipitation process. Process trajectory was constructed based on principal component analysis (PCA) of the real-time PAT data, including process NIR spectra and process turbidity data. The process trajectory clearly reveals various distinguishable process events and process signatures, such as incubation, nucleation, and crystal growth. The practical implication of the process knowledge gained for process control and the establishment of operational process space will be discussed.

**Keywords**: Quality-by-Design (QbD), Process Analytical Technology (PAT), Co-precipitation, real-time process monitoring, incubation, nucleation, crystal growth, Principal Component Analysis, Process Trajectory.

*Corresponding author: Huiquan Wu, Phone: 301-796-0022; Email: Huiquan.wu@fda.hhs.gov

## INTRODUCTION

In the area of pharmaceutical formulation development, co-precipitation of poorly soluble drugs with polymers is an important technique for improving the dissolution and absorption of drugs. This process has been modified in recent years to prepare extended release preparations, for example, naproxen microsphere preparations. Naproxen is a potent non-steroidal anti-inflammatory drug used as the treatment of rheumatoid arthritis, osteoarthritis, acute gout, and as analgesic and antipyretic. However, its use is frequently limited due to significant gastrointestinal side effects. In this study, naproxen was selected as a model drug as it has a short half life, gastrointestinal effects, is soluble in alcohol and practically insoluble in water. Eudragit L 100 was used as it yields microspheres without any need of additives.

Previous work[1-3] included optimization of process variables for the preparation of ibuprofen co-precipitates with Eudragit S100, screening of process and formulation variables for the preparation of extended release Naproxen tablets with Eudragit L 100-55, and preparation and characterization of co-precipitate of ibuprofen using different acrylate polymers. etc. However, little work has been reported regarding the use of an integrated Process Analytical Technology (PAT) approach to gain detailed process information, process knowledge, and process understanding[4] for a co-precipitation process. On the other hand, crystallization as a critical unit operation for active pharmaceutical ingredient (API) production has attracted a lot of attention. Online process analyzers offer unprecedented opportunities to gain detailed process information for illustrating the progress of a multi-phase process and for elaborating the sequential events that take place during the process[5] .However, to the best of our knowledge, practical examples are still lacking for illustrating the feasibility of using PAT tools to monitor various crystallization process events, in terms of providing key process information to distinct various process stages which are essential to both crystallizer design and crystallization process control. In this work, an integrated real-time process monitoring strategy was developed to develop the process trajectory for a pharmaceutical co-precipitation (naproxen and eudragit L100) process.

## EXPERIMENTAL

### Materials

Naproxen USP (lot No: NPX 368) was obtained from Albemarle Corporation (Orangeburg, SC). Eudragit L 100 (lot No: 1221203048) was obtained from Röhm America Inc. (Somerset, NJ). Solvent (A) reagent alcohol (HPLC grade, lot No: 053546) was purchased from Fisher Scientific. Solvent B DI water was obtained from a FDA in-house facility and was kept in refrigerator at $4^0$C prior to use. All of these chemicals and solvents were used without any further processing or purification prior to the use for this experimental work.

### Equipment and Instruments

Process near infrared (NIR) spectra were acquired with a Luminar™ acoustic-optic tunable-filter (AOTF) based NIR process spectrometer (Brimrose Corporation of America, Baltimore, MD), equipped with a transflectance probe. This AOTF NIR process spectrometer has the capability of online monitoring the changes in the concentration of the key components in the co-precipitation system. The acquisition parameters of Brimrose NIR spectroscopy being used in experiment included: the number of spectra average was 50; no background correction was applied; normal scan type; the gain was 2. One spectrum was recorded and saved automatically every six seconds. A 1.0 cm probe extension was attached to the 679B dark field diffuse reflectance/transmission probe for the actual process monitoring.

Online turbidity measurement was made possible by a Model 2100AN Laboratory Turbidimeter (Hach Company, Loveland, Colorado) with a Masterflex® L/S® model 7518-10 digital fluid pump (Cole-Parmer Instrument Company, Chicago, IL). A flexible tubing connection between the co-precipitation vessel and the sample flow-cell of the turbidimeter was made, such that the fluid could be pumped from the co-precipitation vessel to the sample flow-cell of the turbidimeter for online turbidity measurement, then be returned to the co-precipitation vessel in the reverse direction. Prior to its use for this work, the 2100AN turbidimeter was calibrated using formazin stock solution as recommended by the supplier. Data were reported using Nephelometric turbidity units (NTU), which are specifically compliant for a $90^0$ measurement technique.

Pure components of naproxen and eudragit L100 were scanned by FOSS NIRSystems (Foss NIRSystems, Silver Spring, MD). The schematic of the experimental setup was shown in Figure 1.

**Data Analysis Methods**

**Principal Component Analysis (PCA)**

Principal component analysis (PCA) reduces the dimensionality of a data set consisting of a large number of interrelated variables, while retaining as much as possible of the variation present in the data set. For pharmaceutical QbD and PAT applications where process analyzers are installed, engineers and scientists often need to handle a huge amount of interrelated process data to extract critical process information and knowledge for both process and quality control purposes.

With the help of PCA, the following aspects could be examined: in what aspect one sample is different from another, which variables contribute most to this difference, and whether those variables contribute in the same way (i.e., are correlated) or independently of each other. It can also help to detect patterns, and to quantify the amount of useful information, as opposed to noise or meaningless variation, contained in the data set. In this work, plots of a PC versus another (e.g.: PC1 versus PC2), called score plot, are used to depict the process trajectory.

## Singular Points of a Signal

Advances in modern instrumentation, data storage technologies, and recent FDA's Process Analytical Technology (PAT) Initiative have allowed the pharmaceutical sector to collect huge amount of process data which can be used to extract information about the underlying process. One of the key challenges for process chemometrics lies in, it has been noticed that for a time varying signal in a dynamic system, the information content is not homogenously distributed throughout[6]. Some landmarks such as extreme values and shape changes in the data, termed as singular points (SPs), in the process trajectory contain more information about the dynamic behavior than others. Singular points are used to segment the process signal into regions with homogeneous properties. Because SPs have physical meaning such as beginning or ending of a process event, they can be directly used for state identification, process monitoring, and process supervision. Examples of SPs include points of discontinuities, trend changes, and extrema. SPs were used to detect phase shifts during rifamycin B fermentation experiments[7]. In our recent work, SPs were used to determine powder blending process end-point[8]. In this work, SPs are used to detect the onset of nucleation and crystal growth.

## RESULTS AND DISCUSSION

*Process signature identification using online NIR and chemometrical technique*

NIR is capable of monitoring many pharmaceutical unit operations[8-9], in addition to its well-established pharmaceutical applications in quality control and quality assurance areas[10]. NIR spectrum is information rich and it embraces both the physical and chemical information of pharmaceutical materials or dosage forms that were scanned by the NIR spectroscopy[11]. When applied to pharmaceutical unit operations with the help of on-line NIR process sensors, a lot of process-related information may be captured in real-time. For the co-precipitation process being studied in this work, prior to the introducing the non solvent (water) to the co-precipitation vessel, there was a ternary system of API-eudragit-alcohol already present in the vessel. With the gradual addition of the non solvent into the system, thermodynamically we are moving the overall composition point in the four-component system phase diagram from the liquid phase towards the solid (crystal)-liquid equilibrium line; the zone that is bordered by the solid-liquid equilibrium line (the solubility curve) and the meta-stable limit constitutes the meta-stable zone for co-precipitation. Once the solid-liquid equilibrium line is crossed, co-precipitate begins to form and grow. The overall composition point will be located within the solidus region. As demonstrated through the process maps, the co-precipitation process events were followed successfully. To enhance the resolution of the methodology to be developed for identifying process signatures, the addition rate of the non solvent into the system was kept slow such that the transition of one process state to another state could be observed clearly. The formulation components used, the amount of the non solvent added and its adding sequence, and significant visual observations or indications from turbidity measurement, were summarized in Table 1. As one example, the process 3-D map for batch started with formulation components of 0.1320 g naproxen and 6.0970 g eudragit L100 was shown in Figure 2.

*PCA results of the NIR spectra: seek process signature from the score plots*

While the 3-D process map is able to provide direct visual evidence about what is going on in the process, the sensitivity and resolution of this kind of 3-D direct process map really depends on a number of factors: (1) the sensitivity and resolution of the instrumentation and the signal/noise ratio of the instrumentation used; (2) the mass-transfer rates in the vicinity of process probe tip and the bulk solution in the vessel; (3) the hydrodynamics of the fluid inside the vessel. There is a physical distance between the process probe tip and the point where the non solvent water is introduced and in contact with the bulk solution. In our study, the magnetic stirring of the solution could limit the possible concentration gradient within the vessel, the time lag between the initial moment when the process event take places locally and the moment when the process probe is able to actually detect is always unavoidable due to the existence of physical distance. When a signal associated with a process event is relatively weak, especially when the process event is at its embryo stage, its appearance on the 3-D process map will not be that obvious and may need other tool to help to identify or magnify. On the other hand, some important questions associated with the 3-D process map may need to be answered with the help of other data analysis techniques. Questions include but are not limited to: (1) what are the implications of the inflection points in the 3-D map? (2) Are there any new phase formation during the process? etc. In this case, other techniques such as chemometrics may play a vital role in terms of identifying process signature associated with process event, and provide essential process information regarding what is going on in the process side. In this work, Principal Component Analysis (PCA) was applied to the process NIR spectra, such that critical process information and knowledge could be extracted.

Applied PCA to the process NIR spectra of the above process batch showed that, two principal components are sufficient to characterize the variability embedded with NIR spectra. PC1 and PC2 are able to account for 91% and 9% of the total variability embedded with the process NIR spectra for this batch. When PC2 score is plotted against PC1 score, there is a sharp transition on its PCA score plot as reflected on the slope of the straight line of PC2 score vs. PC1 sore. The slope is dramatically changed from positive value to negative value. As a matter of fact, the two segments of the process trajectory consisting of PC2 score vs. PC1 score are almost perpendicular to each other. When tracking the sample number on this PCA score plot, it was found that: (1) the process trajectory which has positive slope value corresponding to the process state of transparent solution during the course of adding non-solvent into the ternary system; (2) the sharp transition of the process trajectory corresponding to the period that solution starts to become cloudy when the non-solvent is introduced to the ternary system for the 3$^{rd}$ time; and (3) the process trajectory which has negative slope value corresponding to the process state of cloudy solution during the course of adding non-solvent into the ternary system.

The process trajectory on the PCA score plot shows that, process samples 173~238 cover the process window where the nucleation and growth takes place.

***On-line Turbidity Monitoring of the Process***

For the process runs being conducted in this work, we also used turbidity measurement to monitor the turbidity change on-line during the process. The results for two process runs were shown in Figures 4(a) and 4(b). As we can see from Fig 4(a), there are several plateaus observed during the course of introducing the non solvent water into the ternary system. Before adding any non-solvent water to the ternary system, the baseline turbidity data of the ternary system were around 0.33~0.45 NTU. The 1st plateau (1.7~2.6 NTU) and the 2nd plateau (50~70 NTU) are corresponding to the 1st and the 2nd addition of 100ml of non-solvent into the ternary system, respectively. The 3rd plateau (140~150 NTU) occurred when the 3rd addition of 100 ml of non-solvent into the ternary system. Given that the turbidity value of 1st plateau of the four-component system is still very close to the baseline turbidity value of the ternary system, most likely it indicates the system remains as transparent solution and no new phase is formed (as confirmed by the on-line process NIR monitoring). The relative low turbidity values for the 2nd turbidity plateau suggest that the process is still at the incubation stage. At this early stage, the composition point of the system is still located within the liquid region. Therefore, no nuclei were formed and the solution was still transparent and thus had a small NTU value. When more non-solvent was introduced gradually, the 3rd turbidity plateau (140~150 NTU) occurred which is probably correlating to the nucleation process stage. At this stage, the system composition point reaches the solid-liquid equilibrium line and thus nuclei were initiated and a bit higher turbidity values than the 2nd stage was detected. Afterwards, when more non-solvent was introduced, the system composition point reaches the meta-stable zone. Consequently, a large super-saturation was created and a crystal growth period was observed. This stage has a turbidity plateau of 2360~2770 NTU. Whit more and more non-solvent was introduced, the system approaches a quasi-steady state where an even higher turbidity plateau of 4000~5180 NTU was observed.

Turbidity profile for a formulation with a lower drug-loading (0.06942g API and 5.96208g Eudragit L100) was acquired in a similar fashion and was shown in Figure 4(b). Before adding any non-solvent water to the ternary system, the baseline turbidity data of the ternary system were around 1.67~2.56 NTU. The 1$^{st}$ plateau (30~50 NTU) and the 2$^{nd}$ plateau (94~195 NTU) are corresponding to the 1$^{st}$ and the 2$^{nd}$ addition of 100ml of non-solvent into the ternary system, respectively. The 3$^{rd}$ plateau (6028~6550 NTU) occurred when the 3$^{rd}$ addition of 100 ml of non-solvent into the ternary system. Since during the 1$^{st}$ addition of non-solvent to the ternary system, the solution becomes slightly cloudy initially then becomes clear, it may suggest that there were some small nuclei formed initially in the vicinity of the local area of water being introduced into the ternary system and being in contact with the bulk solution due to local building-up of super-saturation, but then the newly formed local nuclei were dissolved quickly due to the de-supersaturation phenomena via mass transfer (diffusion and convection) enhanced by the magnetic stirring of the slurry solution. However, since the existence of fine nuclei was so short and did not have sufficient time to travel to the area which was directly exposed to the Process NIR probe tip, this short shelf-life nuclei was not detected by the Brimrose Process NIRS system. The relative high turbidity values for the 2$^{nd}$ turbidity plateau (94~195 NTU) plus the visible slightly cloudiness of the solution suggest that the process is at the nucleation stage. At this stage, the system composition point reaches the solid-liquid equilibrium line and thus nuclei were initiated and a bit higher turbidity values than the 2$^{nd}$ stage was detected. When more non-solvent was introduced gradually, the 3$^{rd}$ turbidity plateau (6028~6550 NTU) occurred which is probably correlating to a stage which both nucleation and crystal growth are involved, due to the system composition point reaches the meta-stable zone. Afterwards, when more non-solvent was introduced, consequentially, a large super-saturation was created and a crystal growth period was observed.

*Apply PCA to the turbidity data to seek process signature*

To seek additional evidence of process signature from the turbidity profile, PCA was applied to the dataset that consists of the turbidity profiles shown in Figures 4(a) and 4(b). It was found that two principal components (PC1 and PC2) are sufficient to account for the total variance in the dataset: PC1 and PC2 explain 82% and 18% of the total variance, respectively. The PC score plot as shown in Figure 4(c) highlights the process trajectory. It was interesting to note that, the process trajectory follows two straight lines which are perpendicular to each other. The segment that has negative PC1 score falls into one straight line, corresponding to time points 1~84. The other segment that has positive PC1 score falls into another straight line, corresponding to time points 85~115. These two straight lines are perpendicular to each other. Furthermore, in this process trajectory the time point that has positive PC1 score and has closest distance to the negative PC1 score line is 85. By comparing this information with the turbidity profiles shown in Figures 4(a) and 4(b), it can be easily appreciated that this dramatic shift on the process trajectory actually corresponds to the period that a suddenly dramatic increase of the turbidity value, which we deem as indicator of process signature that crystal growth takes place.

## CONCLUSIONS

This work provides a case study that focuses on a process for pharmaceutical PAT application. To the best of our knowledge, it is the first report of using an integrated PAT approach to understand a pharmaceutical co-precipitation process, and to visualize this technologically important process in a 3-dimensional map. The process trajectory constructed based on results of applying principal component analysis (PCA) to either process NIR spectra data or process turbidity profile, clearly demonstrated that various distinguishable process events, including incubation, nucleation, and crystal growth, could be accurately tracked and differentiated. This information and knowledge are essential for developing a suitable design space[12-13] and operational process space for a pharmaceutical unit operation.


## ACKNOWLEDGMENTS

Dr. San Kiang (Process R& D director and technical fellow) at the Pharmaceutical Research Institute of the Bristol-Myers Squibb Company (New Brunswick, NJ), Mr. Steve Ware and Mr. Brendan Simon at the Chemglass Inc. (Vineland, NJ), are acknowledged for their technical discussions on our crystallization system.


## Disclaimer

*The views and opinions expressed in this paper are only those of the authors, and do not necessarily reflect the views or policies of the FDA.*

## REFERENCES


1. Khan MA, *et al.* 1994. Optimization of process variables for the preparation of ibuprofen coprecipitates with Eudragit S100.Int J Pharm 102:185-192.

2. Zaghloul AA, Faltinek J, Vaithiyalingam SR, Reddy IK, and Khan MA. 2001.Naproxen-Eudragit microspheres : screening of process and formulation variables for the preparation of extended release tablets, Pharmazie 56(4):321-324.

3. Kislalioglu M S, Khan M A, Blount C, Goettsch R W, Bolton S. 1991. Physical characterization and dissolution properties of ibuprofen: Eudragit coprecipitates. J. Pharm. Sci..80 :799-804.

4. FDA, 2004. "Guidance for Industry. PAT—A Framework for Innovative Pharmaceutical Development, Manufacturing, and Quality Assurance". Available at: http://www.fda.gov/cder/guidance/6419fnl.pdf

5. Caillet A, Puel F, and Fevotte G. 2008. Quantitative in situ monitoring of citric acid phase transition in water using Raman spectroscopy. Chemical Engineering and Processing 47: 377-382.

6. Srinivasan R, Qian MS, 2005. Off-line temporal signal comparison using singular points augmented time warping. Ind. Eng. Chem. Res. 44: 4697-4716.

7. Doan X-T, Srinivasan R., Bapat P. M., Wangikar P. P., 2007. Detection of phase shifts in batch fermentation via statistical analysis of the online measurements: A case study with rifamycin B fermentation. J. Biotechnology, 132:156-166.

8. Wu H and Khan MA, 2009. Quality-by-Design (QbD): An Integrated Approach for Evaluation of Powder Blending Process Kinetics and Determination of Powder Blending End-Point. J Pharm Sci. 98(8):2784-98.

9. Sulub Y, Wabuyele B, Gargiulo P, Pazdan J, Cheney J, Berry J, Gupta A, Shah R, Wu H, and Khan M. 2009. Real-time On-line Blend Uniformity Monitoring Using Near-Infrared Reflectance Spectroscopy: A Noninvasive Off-line Calibration Approach. J. Pharm.l Biomed. Anal. 49:48-54.

10. Ciurczak EW and Drennen JD III. "Near-infrared spectroscopy in pharmaceutical applications," in *Handbook of Near-Infrared Analysis*, 2nd edition, edited by Burns DA and Ciurczak EW, Marcel Dekker, New York, 2001

11. Wu H, Lyon RC, Hussain AS, Drennen JD III, DasGupta D, Voytilla RJ, Bejic L. 2003. Application of Principal Component Analysis in Assessing Pharmaceutical Formulation Design: Exploring the Casual Links between the Tablet Processing Conditions and Drug Dissolution Rate. In: Proceedings of Symposium on Innovations in Pharmaceutical and Biotechnology Development and Manufacturing, 2003 AIChE Annual Meeting [TE001], pg. 635-642. San Francisco, Nov. 17-21, 2003.

12. FDA/ICH, 2006. Guidance for Industry. Q8 Pharmaceutical Development. Available at: http://www.fda.gov/cder/guidance/6746fnl.pdf

13. FDA/ICH, 2008. Q8(R1) Pharmaceutical Development Revision 1. Draft available at: http://www.fda.gov/cder/guidance/8084dft.pdf

Table 1. Formulations used, process stream introduced, and subsequent process events observed via process NIR system, visual observation for the co-precipitation system during the course of adding the non solvent (B) into the ternary system of (API & Eudragit L100 & alcohol) to form a four-component co-precipitation system

| API (g) | Eudragit L100 (g) | Solvent (A) (ml) | Solvent (B) | | Process events as evidenced from online process monitoring | | |
|---|---|---|---|---|---|---|---|
| | | | Adding sequence | Amount added (ml) | Visual observing | Turbidity | NIR |
| 0.132 | 6.09708 | 280 | 1st | 100 | clear | Not measured | No new phase formed |
| | | | 2nd | 100 | clear | Not measured | No new phase formed |
| | | | 3rd | 100 | clear | Not measured | No new phase formed |
| | | | 4th | 50 | cloudy | Not measured | New phase formed |
| | | | 5th | 50 | cloudy | Not measured | New phase formed |
| | | | 6th | 50 | cloudy | Not measured | New phase formed |
| | | | 7th | 50 | cloudy | Not measured | New phase formed |

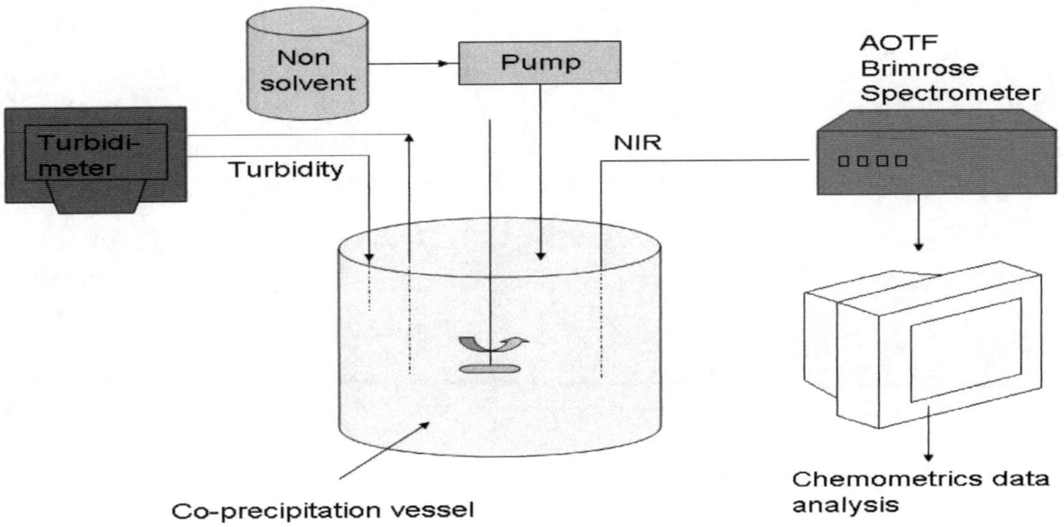


Figure 1 Schematic diagram of experimental setup and process flow with an integrated PAT monitoring system for pharmaceutical API/polymer co-precipitation process using in-line process NIR and on-line turbidity measurement simultaneously.

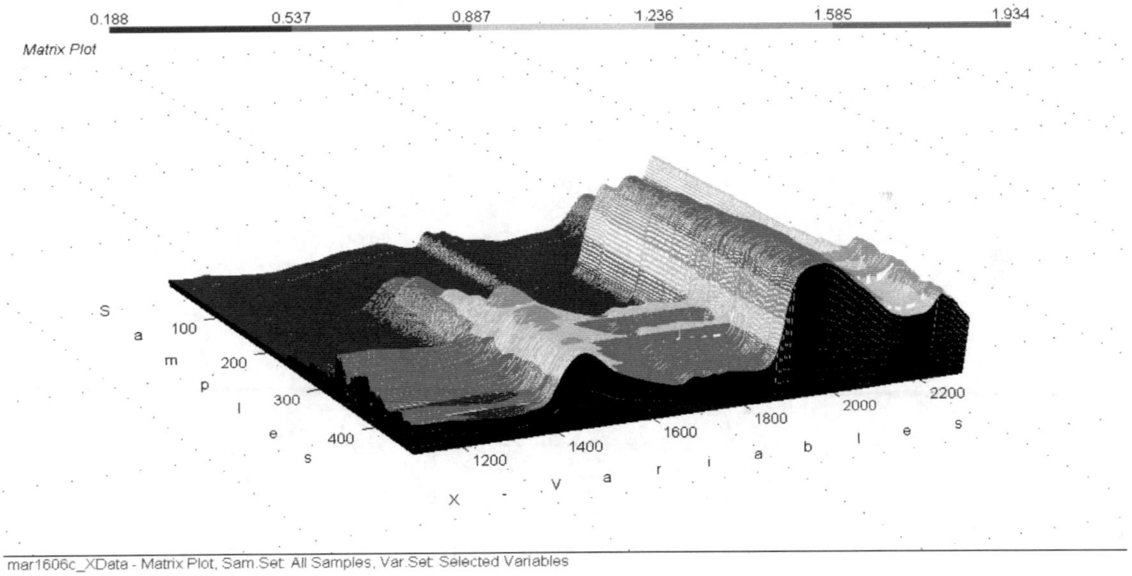


Figure 2 Three-dimensional map of the process state over the entire process course of adding water to the ternary system. An inflection point occurs among process spectra window between spectral number #130~#140. New phase is detected between spectral number #325-335 which maybe the indication of nucleation and growth.

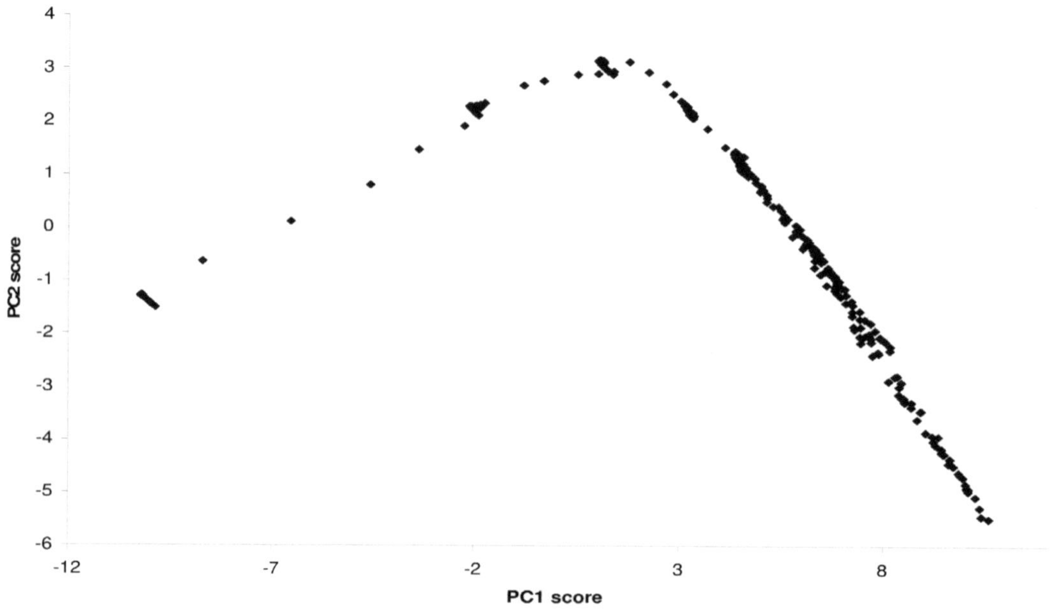


Fig 3: PCA score plot for process NIR spectra of the process bacth started with formulation components of 0.1320 g naproxen and 6.0970 g eudragit L100.

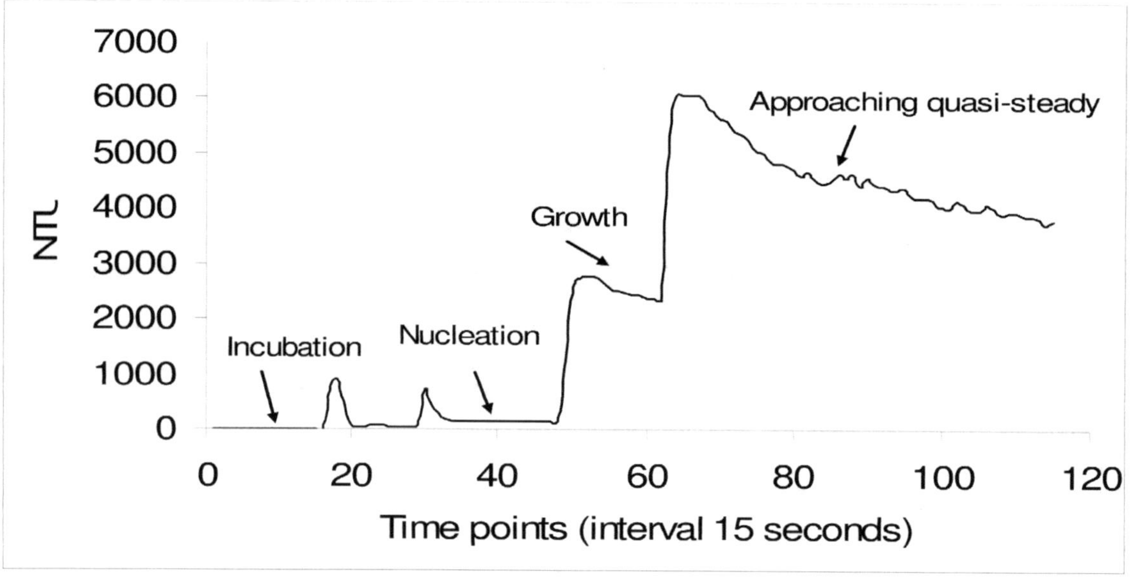


Fig 4 (a) Turbidity profile during the process of sequentially adding water to the ternary system of API (0.1320g), eudragit L100 (6.0970g), and alcohol (280ml) to form the co-precipitate

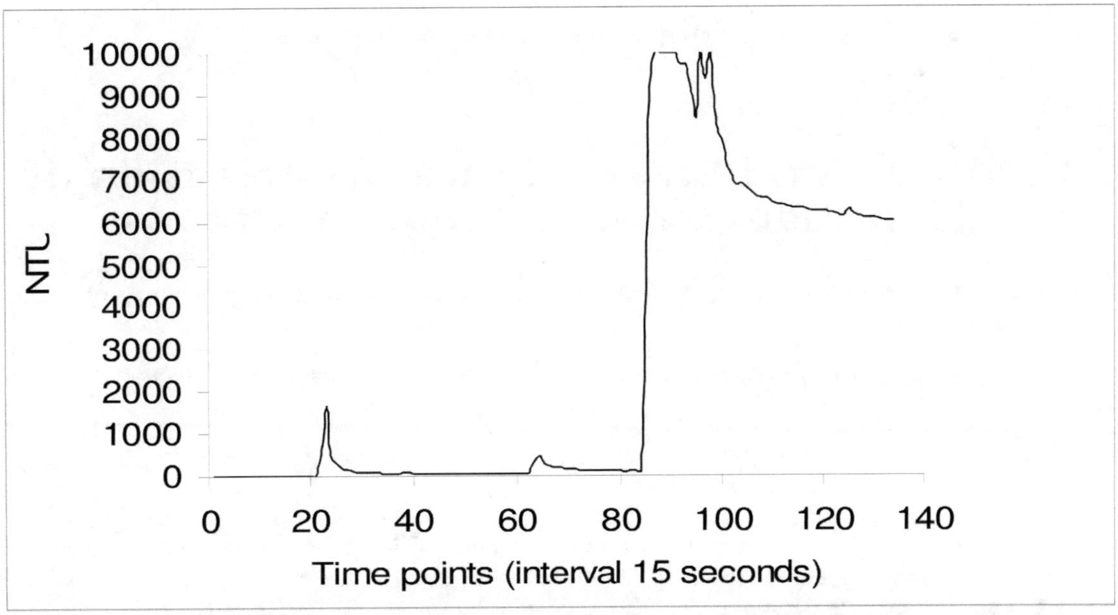


Fig 4(b) Turbidity profile during the process of sequentially adding water to the ternary system of API (0.06942g), eudragit L100 (5.96208g), and alcohol (230ml) to form the co-precipitate

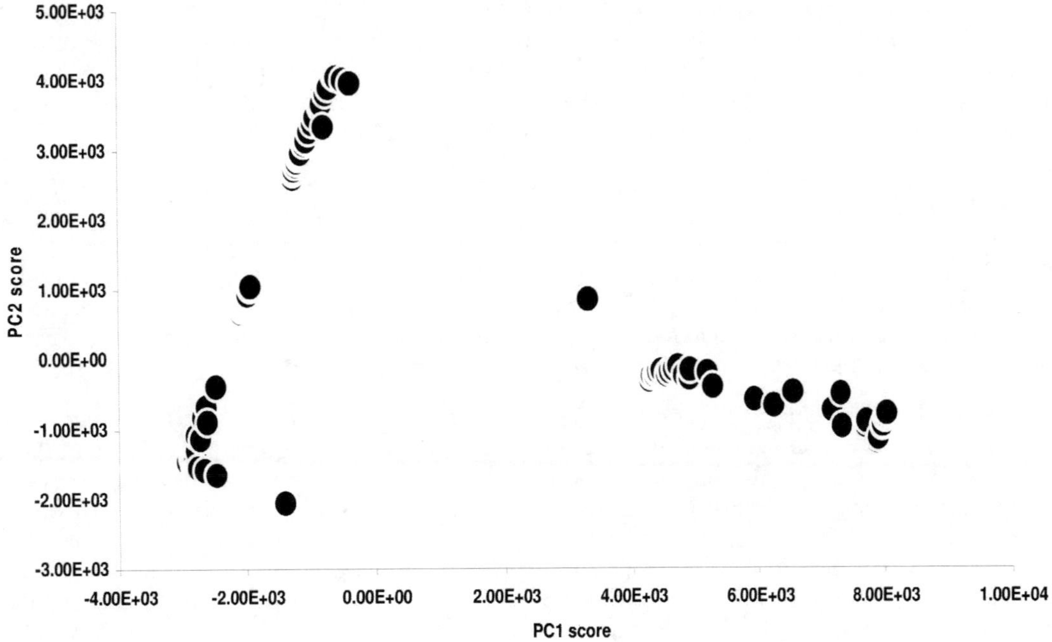


Figure 4(c ). PCA score plot for data set of the real time turbidity profile vs. time (showing that the process trajectory and the dramatic change associated with the crystallization process events)

# Synthesis and Steady-State Optimization of a Continuous Pharmaceutical Process

**Dimitrios I. Gerogiorgis**, Department of Chemical Engineering, Massachusetts Institute of Technology (M.I.T.), Cambridge, MA

Paul I. Barton, Department of Chemical Engineering, Massachusetts Institute of Technology (M.I.T.), Cambridge, MA

*Continuous Pharmaceutical Manufacturing (CPM)* attracts ever-increasing attention nowadays, because of the expanding profitability gap experienced by most pharmaceutical companies; the latter is due to increasing R&D and operating costs and decreasing drug prices (Behr, 2004). Despite the strict licensing requirements for product purity and stability, and the established batch manufacturing processes, the FDA has spearheaded CPM initiatives, recognizing that CPM has a strong potential to improve product quality and suppress operating costs (Plumb, 2005). Comparative studies have confirmed this potential experimentally (Betz et al., 2003). Corporations are thus encouraged to focus on introducing modern process engineering design principles, advances in instrumentation and control, and data management systems (FDA, 2004). The vast majority of *Active Pharmaceutical Ingredients (APIs)* are currently produced by batch processes, whose reconfiguration implies long, expensive clinical trials and revalidation. Nevertheless, novel APIs provide ample opportunity to explore the CPM paradigm, addressing systematically the optimal design and operation of continuous pharmaceutical production lines. *Microsystems technology* offers a convenient, efficient, cost-effective way to conduct a wide variety of organic synthesis reactions (Jensen, 2001) and ensuing separations (Kralj et al., 2007); its advantages include miniaturization (space and fixed investment cost savings), high heat and mass transfer coefficients, scaleability, quick screening, rapid manufacturing and customization. Accelerating CPM R&D (from laboratory to pilot-plant and production scale) can be substantial.

*Organic synthesis* is the cornerstone of CPM process development and flowsheeting, since demonstrating a robust and efficient chemical reaction pathway is essential for proof of concept. In contrast to standard batch process prerequisites (high conversion, yield and API purity), CPM processes can in principle be flexible and viable at much lower conversions and yields, as long as (a) continuous flow without solid precipitation and (b) efficient separations are both ensured. Batch pharmaceutical manufacturing often relies on conventional, recipe-oriented R&D, conducted in long campaigns of arduous, expensive and time-consuming laboratory experiments.

*Advances in unit operations engineering* can also positively impact CPM processes, by improving efficiency as well as reducing solvent emissions and environmental footprint. Nanofiltration (Lin & Livingston, 2007), steady-state recycle (Kennedy et al., 2004) and continuous granulation (Vervaet et al., 2005) are examples of proven technologies which have been shown to enhance the *upstream* (API production) and *downstream* (drug formulation) part of many pharmaceutical production lines, accelerating the transition to CPM processes. Concurrently, breakthroughs in Process Analytical Technology (PAT) (Munson et al., 2006), novel sensors (React-IR) and actuators (micro-pumps) provide in-depth understanding of chemical kinetics, enabling first-principles modeling and efficient multivariate process control.

The present study focuses on plantwide synthesis and optimization of a new upstream process which is aimed at the continuous production of a novel Active Pharmaceutical Ingredient (API). The conceptual flowsheet comprises 3 organic synthesis reactions and 3 separations, considering microsystems (microreactors, microseparators) for the respective steps, due to the plant capacity. Kinetic parameter estimation (based on yield data from CPM experiments conducted at MIT) and thermodynamic modeling enable accurate reactor and LLE extractor design, respectively. Conceptual synthesis allows for solvent integration (Ahmad & Barton, 1999); for costly solvents, this can improve process economics, in contrast to wasteful and exensive batch processing. Steady-state upstream process optimization allows the determination of optimal stream flows and liquid-liquid extraction solvent utilization, based on a number of explicit process design constraints: plant capacity (API throughput), effluent purity, and residual solvent (EMEA, 2001). Cost estimation indicates there is an appreciable margin for cost reduction over batch processes.

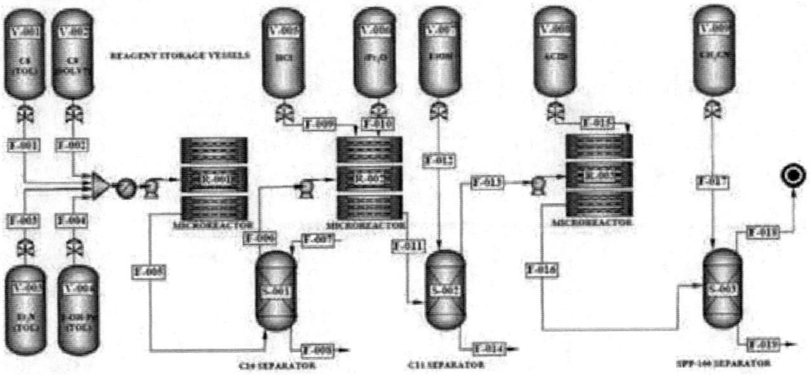

***Figure.*** *The proposed upstream flowsheet of a novel continuous pharmaceutical process for API production.*

**REFERENCES**

Ahmad, B.S., Barton, P.I., Process-wide integration of solvent mixtures, *Comput. Chem. Eng.* **23**(10): 1365-1380 (1999).

Behr, A., New developments in chemical engineering for the production of drug substances, *Eng. Life Sci.* **4**(1): 15-24 (2004).

Betz, G., Junker-Purgin, P., Leuenberger, H., Batch and continuous processing in the production of pharmaceutical granules, *Pharm. Dev. Technol.* **8**(3): 289-297 (2003).

European Medicines Agency (EMEA), *Impurities: Residual solvents*, VICH Topic Report GL18 (plus annexes) (Web: http://www.emea.europa.eu/pdfs/vet/vich/050299en.pdf) (2001).

U.S. Food and Drug Administration (FDA-CDER), *Innovation & continuous improvement in pharmaceutical manufacturing.* (Web: www.fda.gov/cder/gmp/gmp2004/manufSciWP.pdf) (2008).

Jensen, K.F., Microreaction engineering: is small better?, *Chem. Eng. Sci.* **56**(2): 293-303 (2001).

Kennedy, J.H., Belvo, M.D., Sharp, V.S., Williams, J.D., Comparison of separation efficiency of early phase active pharmaceutical intermediates by steady state recycle (SSR) and batch chromatographic techniques, *J. Chromatogr. A* ***1046***(1-2): 55-60 (2004).

Kralj, J.G., Sahoo, H.R., Jensen, K.F., Integrated continuous microfluidic liquid-liquid extraction, *Lab Chip* **7**: 256-263 (2007).

Lin, J.C.T., Livingston, A.G., Nanofiltration membrane cascade for continuous solvent exchange, *Chem. Eng. Sci.* **62**(10): 2728-2736 (2007).

Munson, J., Stanfield, C.F., Gujral, B., A review of process analytical technology (PAT) in the US pharmaceutical industry, *Curr. Pharm. Anal.* **2**(4): 405-414 (2006).

Plumb, K., Continuous processing in the pharmaceutical industry - Changing the mind set, *Chem. Eng. Res. Des.* **83**(A6): 730-738 (2005).

Vervaet, C., Remon, J.P., Continuous granulation in the pharmaceutical industry, *Chem. Eng. Sci.*
Abstract: Synthesis and Steady-State Optimization of a Continuous Pharmaceutical Process (2009 Annual Meeting)
**60**(14): 3949-3957 (2005).

See more of this Session: Qbd Across Drug Substance Operations

See more of this Group/Topical: Topical I: Comprehensive Quality by Design in Pharmaceutical Development and Manufacture

# Design Space Continuum: From Drug Substance to Drug Product

**T. Kourti**, GlaxoSmithKline and McMaster University, Hamilton, ON, Canada

Recognising that drug production is a continuum that spans from Drug Substance to Drug Product, will result in the development of a more versatile and robust Design Space.

The final product that delivers the active pharmaceutical ingredient to a patient, is indeed a multidimensional combination of the raw material attributes and the process parameters of several unit operations including those of the drug substance production (like reaction and crystallization), the ones from drug product production (like granulation and compression) but also packaging. Each one of the unit operations has an impact on one or more final product quality characteristics like dissolution and uniformity, but also on stability.

Therefore, if the Design Space is addressed in a holistic form, then all raw material and unit operations involved in the production and their interactions should be considered.

See more of this Session: Qbd Across Drug Product Operations
See more of this Group/Topical: Topical I: Comprehensive Quality by Design in Pharmaceutical Development and Manufacture

# Implementing Risk- and Knowledge Management Tools into Product and Process Development According to Quality by Design

Brigitte Guebitz[1], Johannes Fritz[2], Hubert Schnedl[2], Siegfried Adam[1], Stefan Radl[1], Johannes Khinast[3]

[1]) Research Center Pharmaceutical Engineering GmbH, Graz, Austria
[2]) THE LIFECYCLE INTELLIGENCE GROUP GmbH, Graz, Austria
[3]) Institute for Process and Particle Engineering, Graz University of Technology, Graz, Austria

By using Quality by Design (QbD) and Process Analytical Technology (PAT) approaches in pharmaceutical development more and even more complex information is generated, which has to be available, managed and shared over the whole development and product lifecycle. Therefore the use of knowledge management tools is becoming an integral part to fulfill the requirements of ICH Q8/9/10.

Ontologies as knowledge management tools are already in use in other producing industries such as automotive and aerospace to manage and exchange information and data in a consistent and transparent way between selected business units across the product lifecycle.
In this paper the combination of the knowledge management tool ontology with the entire quality risk management process is discussed to implement a knowledge and risk based approach into the pharmaceutical product and process development lifecycle according to ICH principles.

In this first step, the ontology is seen as a domain-specific structured knowledge base for 'Failure Mode and Effect Analysis' (FMEA) and 'Cause-and-Effect Diagram' (CED) related concepts and entities. Therefore objects of different domains such as risk analysis process, requirements, organization and others are modeled according to the newly developed 'GxP-OntoV' methodology.

This new methodology extends traditional ontology development processes to fulfill the needs of GAMP 5.0 and GxP requirements for computerized systems. An interdisciplinary approach was created, so that the ontology is specified comprehensively from the beginning and can be properly verified and successfully validated.

As development environment the open source ontology editor 'Protégé' is used. The open standard 'Web Ontology Language' (OWL) serves as implementation language.
Subsequently the ontology is extended by integration of further related concepts and entities e.g. product and process development steps, validation activities.

For the presentation the 'GxP-OntoV' methodology as well as the first development results of the 'Quality by Design/Risk Management Ontology' will be discussed in detail.

# Application of QbD Principles to Analytical Methods

**Kimber Rundlett**, Analytical Research & Development, Pfizer Global Research & Development, Groton, CT

The same QbD Principles applied to the design of manufacturing processes are applicable to the design of analytical methodology. QbD elements such as understanding the science behind the methods, and improving method reliability by identifying and controlling sources of variability are key. This presentation will describe how analytical methods can be developed, validated and transferred by following QbD principles and how analytical method design space can be established. Advantages of applying QbD to analytical methodology include development of more robust methods, better understanding of analytical methods and their variability, facilitation of manufacturing process improvements and potential.regulatory flexibility allowing changes to methodology within the established design space.

See more of this Session: Qbd Across Drug Product Operations
See more of this Group/Topical: Topical I: Comprehensive Quality by Design in Pharmaceutical Development and Manufacture

# The Development of Methods to Link Design Space Models to Product Stability

Zhixin Zong, Division of Pharmaceutics, The University of Iowa, Iowa City, IA

Peter L. Wildfong, Pharmaceutics, Duquesne University, Pittsburgh, PA

Lee E. Kirsch, Division of Pharmaceutics, The University of Iowa, Iowa City, IA

Aditya M. Kaushal, Department of Pharmaceutics, University of Minnesota, Minneapolis, MN

Eric J. Munson, Pharmaceutical Chemistry, University of Kansas, Lawrence, KS

Raj Suryanarayanan, Department of Pharmaceutics, University of Minnesota, Minneapolis, MN

Dewey H. Barich, Pharmaceutical Chemistry, University of Kansas, Lawrence, KS

Kalyana Pingali, Department of Chemical and Biochemical Engineering, Rutgers University, Piscataway, NJ

Fernando J. Muzzio, Department of Chemical and Biochemical Engineering, Rutgers University, Piscataway, NJ

Defne Kayrak-Talay, Chemical Engineering, Purdue University, West Lafayette, IN

Salil D. Desai, Division of Pharmaceutics, The University of Iowa, Iowa City, IA

Ira Shea Buckner, Pharmaceutics, Duquesne University, Pittsburgh, PA

Members of the National Institute for Pharmaceutical Technology and Education (NIPTE) have been working on an FDA-sponsored project entitled: "Development of Quality by Design (QbD) Guidance Elements on Design Specifications Across Scales with Stability Considerations". The overall objectives of the project are to improve pharmaceutical product quality and maximize process innovation and continuous quality improvements by: a) developing QbD guidance elements on process design space, scale-up, and process validation for three unit operations: mixing, granulation, and drying; and b) developing a framework for optimizing design space specification across scales with consideration to stability.

The key to linking design space models to product instability is the identification of a linkage between physical change in API that result in a changes in the susceptibility of that API to chemical degradation. The presence of this linkage means that monitoring the physical state of the API may predict the product's susceptibility of chemical instability failure.

The model drug for this project is gabapentin which is known to degrade to gabapentin lactam. The degradation is sensitive to formulation and processing conditions. The topic of this presentation is an interim report on the development of methods and the assembly of correlations between physical methods and susceptibility of chemical instability based on the evaluation of materials generated in laboratory scale process simulations.

Our approach has been to 1) screen a series of potentially using solid-state characterization methods and to install and validation stability-indicating chromatographic methods and 2) to subject the various physical forms of the API and powder blends to typical processing and storage stresses to determine whether quantifiable changes can be observed. Specifically, we have evaluated stability-indicating HPLC, SS-NMR, differential scanning calorimetry, and powder X-ray diffraction as potential characterization and/or stability predictors. Samples of gabapentin were subjected to processing stresses associated with granulation, milling, drying, and blending.

Physical Characterization Methods:

A thermogravimetric analyzer (TGA, Model Q50, TA Instruments) and a differential scanning calorimeter (DSC, Model 2920, TA Instruments, New Castle, DE) were connected to a thermal analysis operating system (Thermal Analyst 2000, TA Instruments). Approximately 2 to 6 mg of the sample, in a crimped aluminum pan or an open aluminum pan, was heated in the DSC/TGA from room temperature to 250 C at varying rates under nitrogen purge. XRD analysis was conducted on samples in a glass holder and exposed to Cu Ká radiation (45 kV x 40 mA) in a wide angle X-ray diffractometer (model D5005, Bruker, Madison, WI) at ambient temperature. The instrument was operated in a step-scan mode, in 0.05°2q increments, and counts were accumulated for 1.0 second at each step over the angular range of 5 to 40°2q. Data analyses were performed with commercially available software (JADE, version 8.0, Materials Data Inc., Livermore, CA). Solid state NMR spectra were acquired on a CMX spectrometer operating at 300 MHz for proton and 75 MHz for carbon. Samples were packed into a Revolution NMR 7 mm (o.d.) zirconia rotor. The spectra were acquired with cross polarization (CP) using a linear ramp on the proton channel during the CP period. The optimal contact time varied between samples, but was typically around 0.9 ms. Magic angle spinning (MAS) was done at a rate of 4.0 kHz. Proton spin-lattice relaxation data were acquired via saturation recovery. Spinning sidebands were eliminated from non-relaxation spectra via sideband suppression techniques.

Chromatographic methods:

RP-HPLC analysis was carried using a high/low assay wherein trace amounts of degradation product (lactam) were measured using high concentrations of parent API (12 mg/mL) and sample API amounts were measured at about 2.4 mg/mL. These two results were then used to compute the lactam content as a percentage (w/w ) of API. Calibration standards for the lactam and gabapentin were prepared in the range of 0.5-5 µg/mL and 0.5 to 5 mg/mL, respectively. The HPLC system was a

Thermo Spectrum System P4000 pump, AS3000 auto injector, and UV 6000 LP photodiode array detection system. The column was ìBondapak Cyano column 3.9x300mm. Isocratic analysis was conducted using a mobile phase composed of 95 parts buffer (10 mM KH2PO4/10 mM K2HPO4 ) and 5 parts acetonitrile . Analysis was carried out using an analytical wavelength of 210 nm, flow rate of 1.0 mL/min and a 0.020 mL injection volume. Run times were 10 minutes. Retention volumes for gabapentin and lactam were 4 and 8.5 mL, respectively. Calibration plots were linear. To evaluate the susceptibility of gabapentin materials to thermally-induced degradation, an aliquot of API (approximately 15 mg ) was weighed and placed in type II glass vial, sealed with a Teflon-faced butyl rubber stopper and crimp-secured with an aluminum seal. Vials were wrapped with aluminum foil, and then stored at 50 °C for 24 hours. A 1.0 mL aliquot of mobile phase was injected to the vial by using 25 G needle and 1.0mL syringe. The vials were shaken until samples dissolved. Solutions were assayed for lactam (high concentration) and diluted 5-fold with mobile phase of analysis of gabapentin.

Preparation of process stress materials:

Samples were subjected to milling using a Planetary Micro Mill "Pulviserette 7" mill from Fritsch GMBH (Idar-Oberstein, Germany) using two forty-five mL chromium hardened steel grinding bowls with four 15 mm (diameter) chromium hardened steel balls. Milling was conducted for 15 to 60 minutes. Spray-dried materials were prepared using a 5% w/w gabapentin solution in water:methanol (1:1 or 1:2 v/v) spray dried using a Buchi spray dryer. Inlet temperature was set between 70 °C, aspirator was set at 100% of capacity, and pump was used at 5% capacity. Freeze dried material was prepared using a 10 % (w/v) solution frozen at -45 °C. Primary drying was done at -30 °C, and secondary drying was done at +20 °C. Additionally a solution of gabapentin, gabapentin:PVPK90 (weight ratio of 6:1) and a suspension of gabapentin:HPC-L (weight ratio of 6:1) were also freeze dried. Granulation samples were prepared from single drop experiments with water or aqueous solutions containing 1 to 5% (w/v) HPC as the binder. Aliquots of these materials were subjected to marumerization using a flat disc marumerizer for 1 minute at 200 rpm. Granulation samples were also generated using a Diosna granulator. Blending samples were prepared using a shear rheometer at 5 rpm,20 rpm and 80 rpm. At each shear rate, the conditions of exposure time(shear strain) were 40, 80, 160, 320, 640 and 1280 revolutions. These shear strain conditions were used for every shear rate. At each stage of shear strain condition, a sample of 1g was collected.

Results: Lactam levels increased by greater than 50-fold for samples subjected to lyophilization and milling. Elevated lactam levels (4 to 20-fold) were also observed in samples after granulation and mixing stress. Susceptibility of processed samples to post-processing instability was evaluated by thermal-stress (50 °C for 24 hours). The initial rate of lactam formation ranged from 0.000060 mole %/hour to 0.04 mole %/hour depending on processing history of the sample. In general, the initial lactam formation rate was proportional to the sample surface area. The presence of expicients (hydropropyl cellulose or polyvinylpyrolidine) increased the initial lactam formation rate. Surprisingly, exposure of samples to high humidity in combination with thermal stress decreased the initial rate of lactam formation. Some processed samples (lyophilized and spray-dried) contained detectable levels of gabapentin hydrate (Form I) although no correlation between in-process lactam formation or post-processing instability could be ascertained. An increase in T1 relaxation time by SS-NMR was observed when samples were processed under rigorous conditions such as milling.

Conclusion: Correlations between in-process lactam formation or T1 relaxation time and post-processing instability were observed. Thus the detection of trace levels of lactam formation during processing (using chromatographic or spectroscopy methods) appears to be a predictive indicator of the susceptibility of gabapentin to post-processing instability.

See more of this Session: Qbd Across Drug Product Operations
See more of this Group/Topical: Topical I: Comprehensive Quality by Design in Pharmaceutical Development and Manufacture

# Establishing Appropriate Particle Size Specifications for Solid Oral Dosage Forms

**Zhigang Sun**, Division of Chemistry II, Office of Generic Drugs, OPS, CDER, FDA, Rockville, MD

Control of particle size distribution of pharmaceutical powders is critical in the manufacture of solid oral dosage forms due to significant impact of particle size on the safety and efficacy of the drug product. As shown in many publications, particle sizes of powders have profound influences on not only drug product performance (e.g., dissolution, content uniformity, and stability etc.) but also drug product manufacturability (e.g., flowability, blend uniformity, compactibility etc.). Therefore, establishing appropriate particle size specifications is necessary for control of drug product quality and manufacturing consistency.

Traditionally, in the manufacture of solid oral dosage forms, control of particle sizes is only considered for the drug substance with low solubility. However, for the Quality by Design (QbD) approach, the effects of particle sizes of the drug substance, excipients, and in-process materials on critical product attributes (CQA) of the drug product should be assessed based upon drug formulation and manufacturing process. If particle sizes of some of raw or in-process materials are identified as the critical material attributes (CMA) during manufacturing processes (e.g., mixing, granulation, milling, blending, coating etc.), the particle size specification is required for each of these materials, which may include not only the drug substance but also excipients and in-process materials. In this presentation, some examples will be discussed to demonstrate when the particle size specification needs to be established based upon the QbD approach.

Once a particle size specification of pharmaceutical powders is required, the next question is how to establish an appropriate particle size specifcation. In many new drug applications (NDAs) and abbreviated new drug applications (ANDAs), it is common to see that the particle size specifications submitted are not complete, not really meaningful, or not adequate to control the particle size distribution of pharmaceutical powders. Therefore, it is important to know what information should be included in the particle size specification to meet the regulatory requirement. In this presentation, for each component of particle size specifications (analytical procedures, method validation, and acceptance criteria), recommendations will be discussed regarding establishment of an appropriate particle size specification.

See more of this Session: Qbd Across Drug Product Operations
See more of this Group/Topical: Topical I: Comprehensive Quality by Design in Pharmaceutical Development and Manufacture

# Using QbD approaches to enable development of real time release for a continuous drug product process

Dave Sharp
Pfizer Global Manufacturing
Sandwich, UK

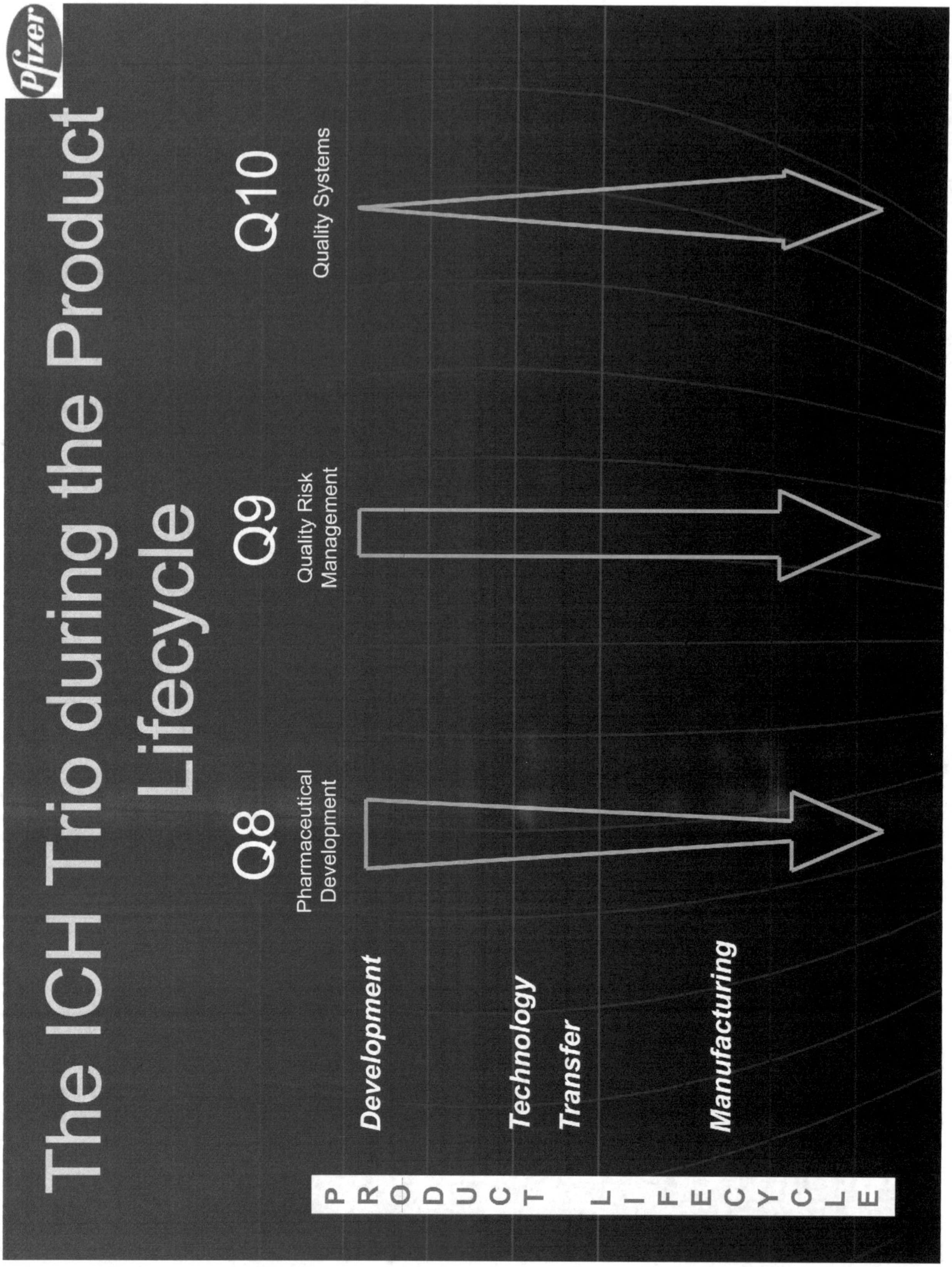
The ICH Trio during the Product Lifecycle
Pfizer
PRODUCT LIFECYCLE
Development
Technology Transfer
Manufacturing
Q8
Pharmaceutical Development
Q9
Quality Risk Management
Q10
Quality Systems

# A Case Study Example

- Pfizer's "Next Generation" manufacturing initiative – an extension of Quality by Design to achieve
  - Continuous manufacturing of a drug product
  - Control Strategy - use Process Analytical Technology (PAT) and Design Space to demonstrate control
  - Process validation by Continuous Quality Verification (CQV)
  - Introduction of Real Time Release (RTR)
- Immediate release solid oral dosage form (capsule)

# Continuous Manufacturing

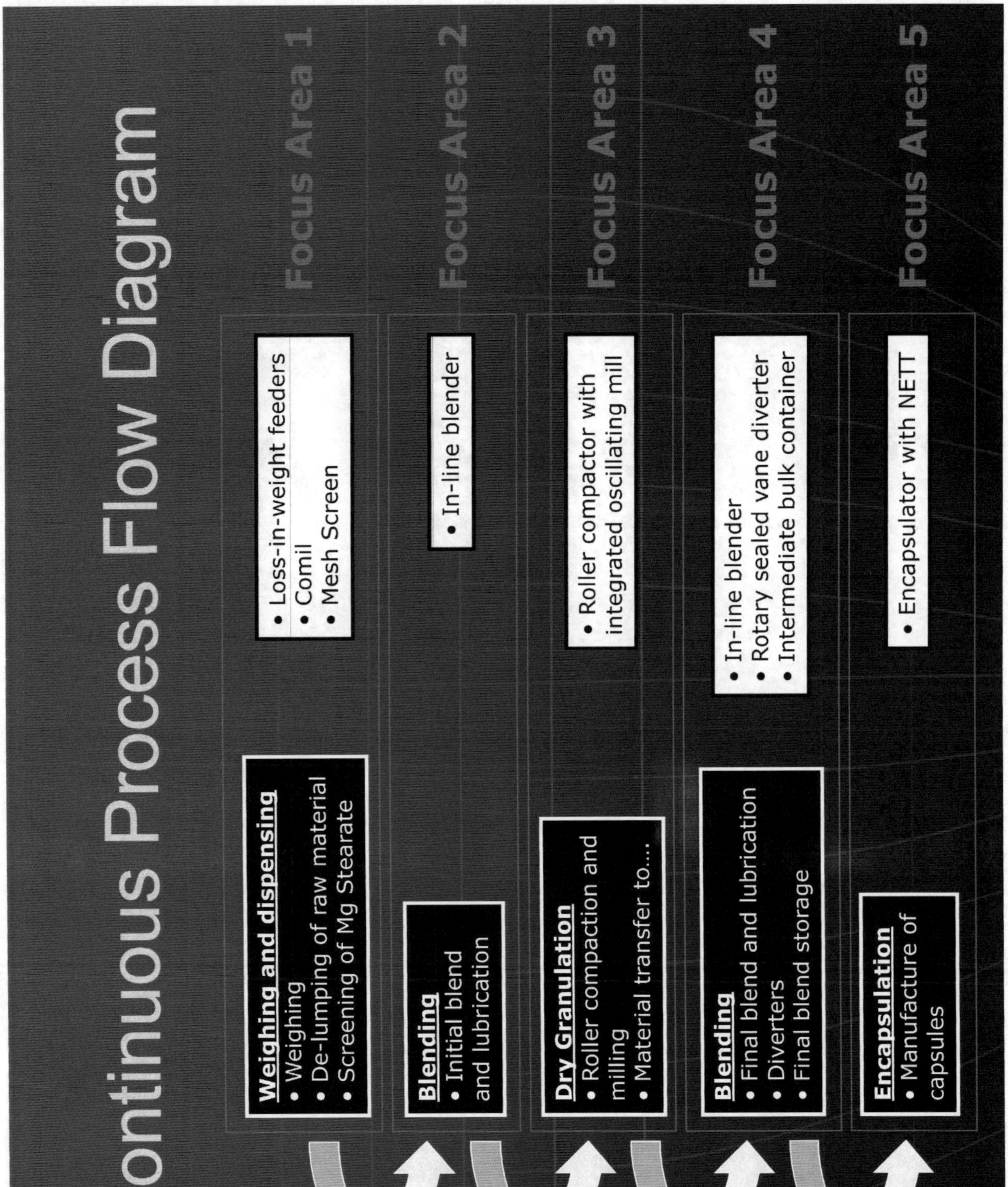

Continuous Process Flow Diagram
Weighing and dispensing
• Weighing
• De-lumping of raw material
• Screening of Mg Stearate
• Loss-in-weight feeders
• Comil
• Mesh Screen
Focus Area 1
Blending
• Initial blend and lubrication
• In-line blender
Focus Area 2
Dry Granulation
• Roller compaction and milling
• Material transfer to....
• Roller compactor with integrated oscillating mill
Focus Area 3
Blending
• Final blend and lubrication
• Diverters
• Final blend storage
• In-line blender
• Rotary sealed vane diverter
• Intermediate bulk container
Focus Area 4
Encapsulation
• Manufacture of capsules
• Encapsulator with NETT
Focus Area 5
SCADA Control using PAT measurements

# Quality by Design

# Quality by Design at Pfizer

- Quality should be built into the product
  - Testing alone cannot be relied on to ensure product quality
- Pharmaceutical Quality by Design (QbD)
  - Designing and developing formulations and manufacturing processes to ensure predefined product quality
  - Understanding and controlling formulation and manufacturing process variables affecting the quality of a drug product

Pharmaceutical Quality = $f$ (Drug Substance, Excipients, and Manufacturing)

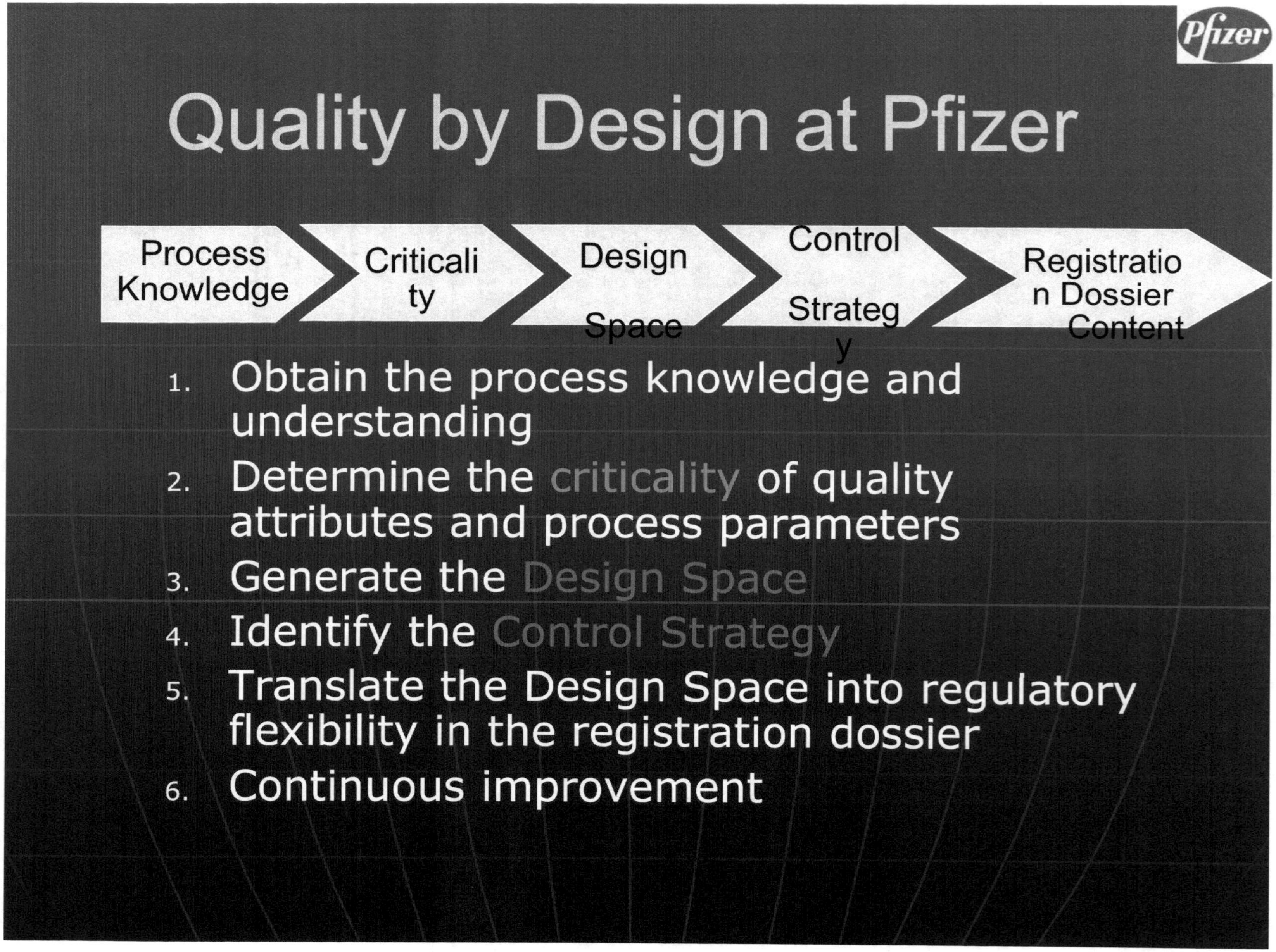
Quality by Design at Pfizer
Process Knowledge
Criticali ty
Design Space
Control Strateg y
Registratio n Dossier Content
1. Obtain the process knowledge and understanding
2. Determine the criticality of quality attributes and process parameters
3. Generate the Design Space
4. Identify the Control Strategy
5. Translate the Design Space into regulatory flexibility in the registration dossier
6. Continuous improvement
Pfizer

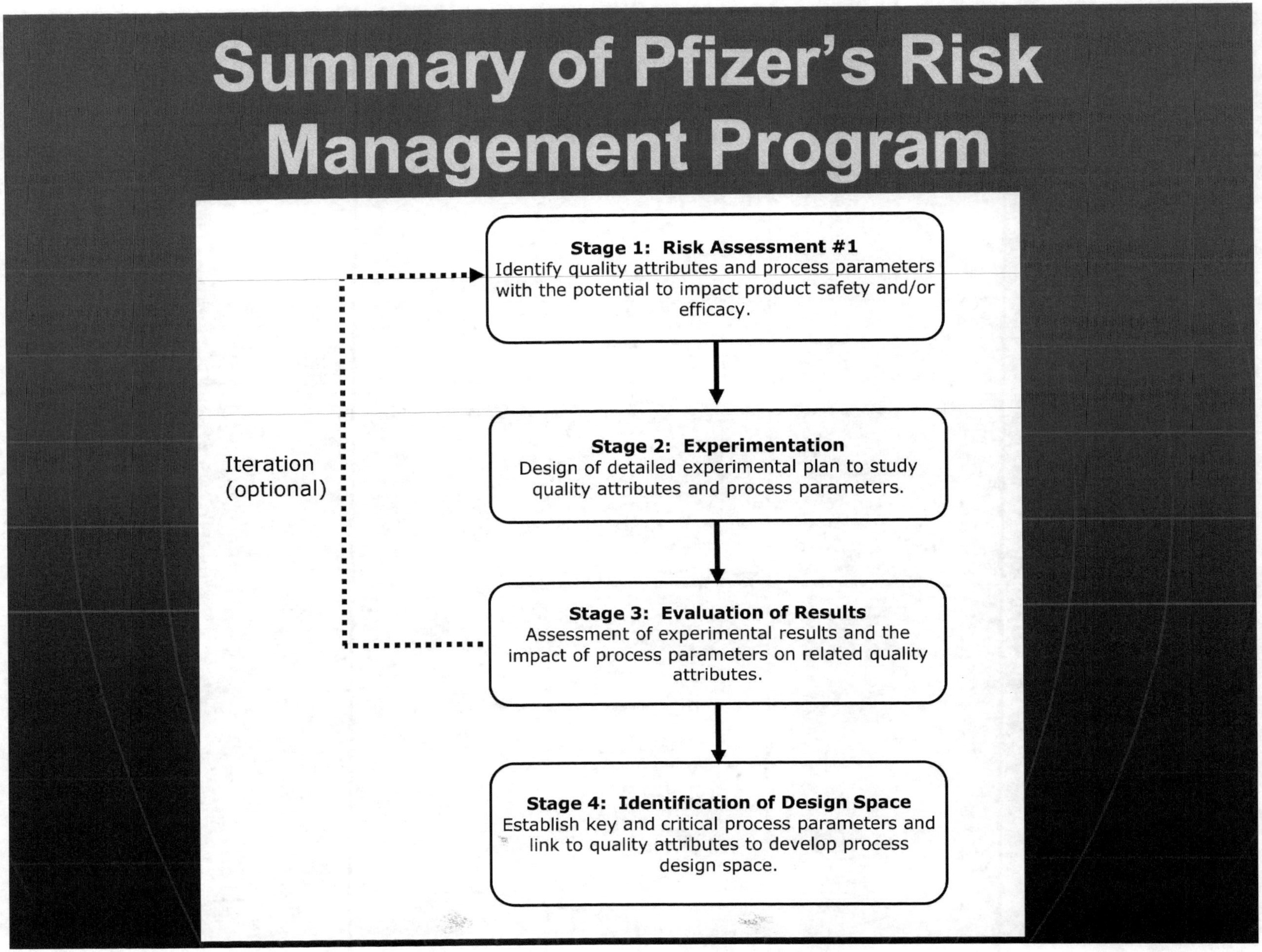
Summary of Pfizer's Risk Management Program
Stage 1: Risk Assessment #1
Identify quality attributes and process parameters with the potential to impact product safety and/or efficacy.
Stage 2: Experimentation
Design of detailed experimental plan to study quality attributes and process parameters.
Stage 3: Evaluation of Results
Assessment of experimental results and the impact of process parameters on related quality attributes.
Stage 4: Identification of Design Space
Establish key and critical process parameters and link to quality attributes to develop process design space.
Iteration (optional)

# Risk Assessment using Cause and Effect Matrices

- Define all relevant Quality Attributes (QA) and Process Parameters (PP)
- Rank importance of QA's then score PP's against each of these in a matrix format
- Prioritise and record rationale
- Define experimental strategy

| Key Attribute Desc | | | | | | | | | | | | | |
|---|---|---|---|---|---|---|---|---|---|---|---|---|---|
| Rank | 1 | 5 | 9 | 10 | 10 | 9 | 5 | 1 | 5 | 9 | | | |
| Parameter | QA1 | QA2 | QA3 | QA4 | QA5 | QA6 | QA7 | QA8 | QA9 | QA10 | Score | Exp't Strategy | Param'r Rationale |
| PP9 | 10 | 10 | 10 | 10 | 1 | 5 | 5 | 5 | 9 | 9 | 461 | | |
| PP5 | 10 | 5 | 10 | 9 | 10 | 10 | 1 | 1 | 1 | 1 | 425 | | |
| PP8 | 1 | 10 | 9 | 1 | 1 | 10 | 5 | 5 | 9 | 9 | 398 | | |
| PP1 | 10 | 10 | 1 | 10 | 9 | 1 | 1 | 10 | 5 | 5 | 353 | | |
| PP10 | 5 | 1 | 10 | 9 | 1 | 1 | 10 | 5 | 5 | 5 | 334 | | |
| PP6 | 1 | 1 | 10 | 9 | 1 | 1 | 10 | 5 | 5 | 5 | 330 | | |
| PP3 | 1 | 1 | 1 | 5 | 10 | 10 | 10 | 5 | 1 | 1 | 324 | | |
| PP4 | 5 | 5 | 1 | 10 | 9 | 1 | 1 | 10 | 5 | 5 | 323 | | |
| PP7 | 5 | 1 | 1 | 5 | 1 | 1 | 10 | 10 | 10 | 10 | 288 | | |
| PP2 | 1 | 10 | 9 | 1 | 1 | 10 | 5 | 5 | 1 | 1 | 286 | | |

# Risk Assessment From Case Study Example

| Parameter | Score | Exp't Strategy | Comments |
|---|---|---|---|
| Equipment maintenance | 590 | FMEA | |
| Inline blender speed (IB) | 520 | DOE | Included in critical DOE |
| Equipment calibration | 470 | FMEA | |
| Stable power supply | 470 | FMEA | |
| Excipient particle size | 410 | DOE | RFT raw material interation DOE |
| API Particle size | 393 | DOE | RFT raw material interation DOE |
| Excipient feed rate | 340 | IA | Have good Ppk data |
| Excipient feed rate factors | 300 | IA | |
| API feed rate | 277 | IA | Have good Ppk data |
| Lubricant feed rate | 277 | IA | Need to improve Ppk's, understand how to do this |
| Excipient moisture content | 275 | DOE | Not studied, can replace with historical data from vendor |
| Lubricant particle size | 267 | DOE | Used surface area in RFT raw material interaction DOE |
| Lubricant surface area | 267 | DOE | RFT raw material interaction DOE |
| API Water Content | 252 | DOE | Can replace with historical data from API site |
| Location of Lubricant feeder | 247 | DOE | Has since been fixed at shortest distance |
| API feed rate factors | 237 | IA | |
| Lubricant feed rate factors | 237 | IA | |
| Lubricant degree of hydration | 217 | IA | |
| Transfer system maintenance | 215 | FMEA | |
| Excipient mill speed | 195 | | Fixed, same as batch process |
| API mill speed | 167 | | Fixed, same as batch process |
| Excipient mill speed | 167 | | Fixed, same as batch process |
| API Potency (LC) | 157 | | Can use historical data to show consistency |
| API Polymorph Confirmation | 147 | | Can use historical data to show consistency |
| API Assay (LC) | 147 | | Can use historical data to show consistency |

# Experimental Strategy

- Two full factorial experimental designs were conducted as part of the development of the continuous manufacturing process.
- Experiments focused on the quality attributes of dissolution and content uniformity.
- Results identified operating ranges for process parameters, and the multivariate approach allowed significant interactions between them to be identified.

# DOE to Design Space

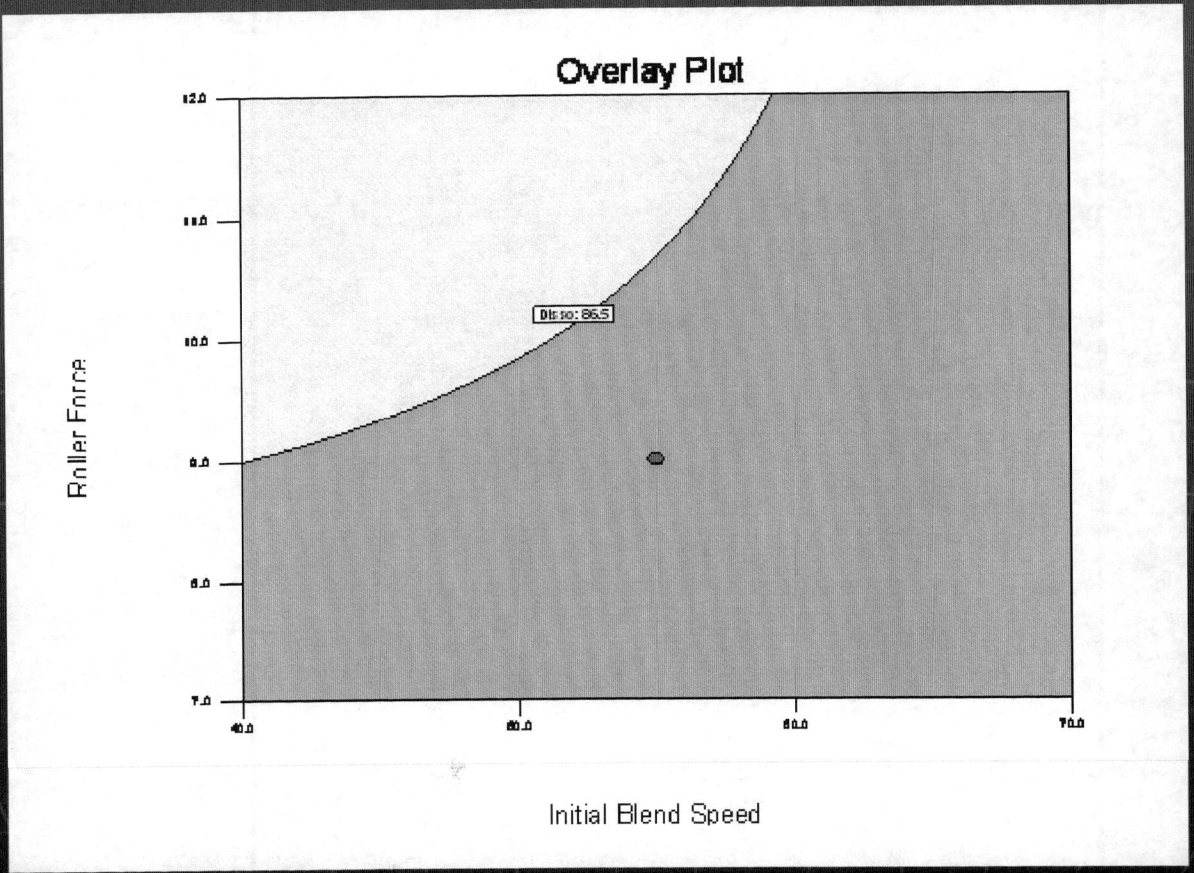

# Design Space and Control Strategy

# Design Space & Control Strategy

- From experiments, define design space:
  - Parameters and inputs → product quality (safety and efficacy)
  - Define ‘critical’, ‘key’, ‘non-critical’ process parameters
  - Identify Proven Acceptable Ranges (PAR)
- Define control strategy
  - Ensures the process operates within the boundaries of the design space and produces acceptable product
  - Includes specifications (raw materials, finished product), process definition and controls, GMP (facilities, equipment)

# Pfizer's Terminology

- Criticality is defined based on a process, rather than trying to identify a specific definition
  - **Critical**: Quality attribute or process parameter that impacts product safety or efficacy
  - **Key**: Process parameter that has the potential to impact product safety or efficacy
  - **Non-Critical**: Quality attribute or process parameter that has no impact on product safety or efficacy

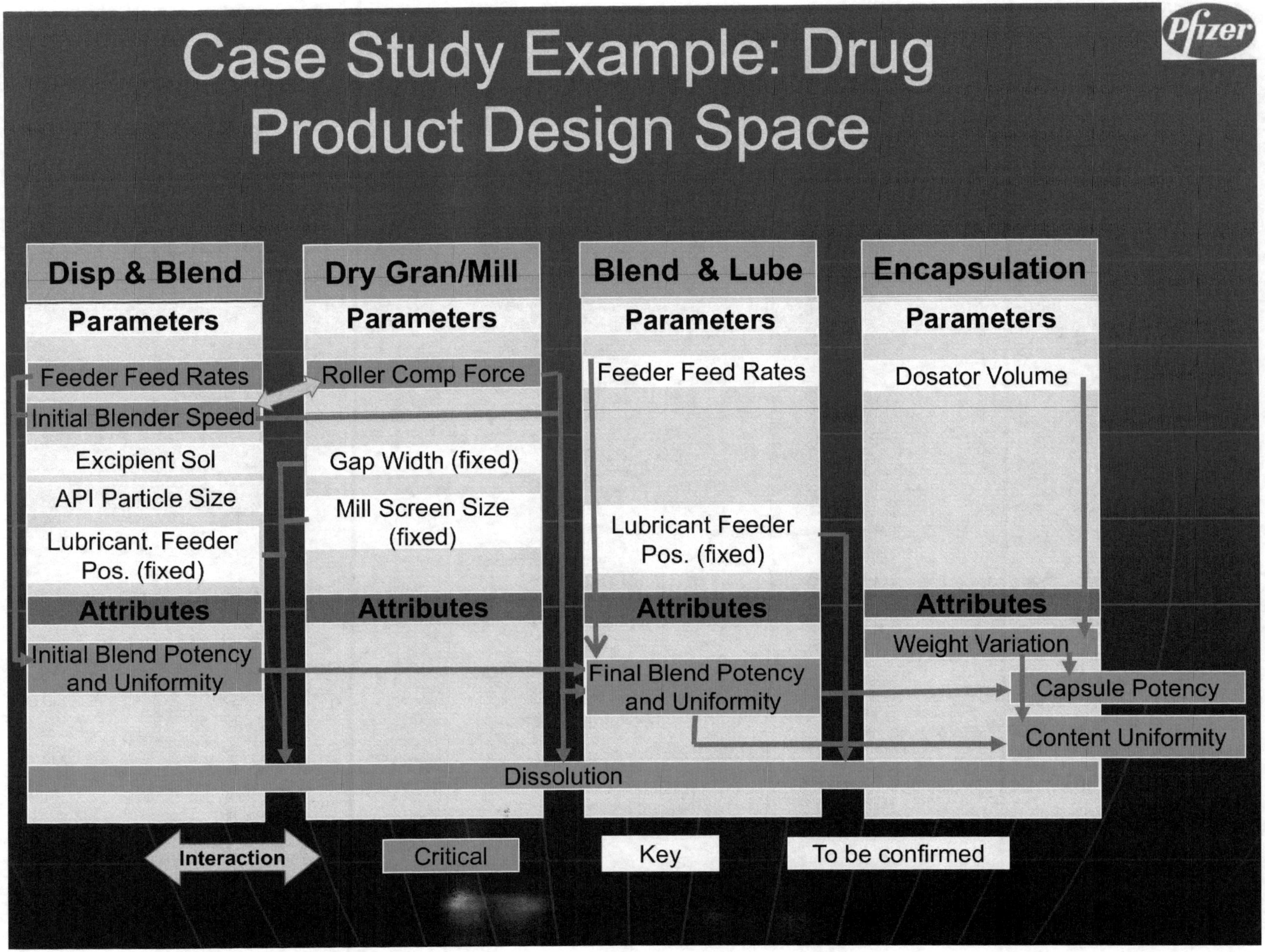
Case Study Example: Drug Product Design Space
Disp & Blend
Parameters
Feeder Feed Rates
Initial Blender Speed
Excipient Sol
API Particle Size
Lubricant. Feeder Pos. (fixed)
Attributes
Initial Blend Potency and Uniformity
Dry Gran/Mill
Parameters
Roller Comp Force
Gap Width (fixed)
Mill Screen Size (fixed)
Attributes
Blend & Lube
Parameters
Feeder Feed Rates
Lubricant Feeder Pos. (fixed)
Attributes
Final Blend Potency and Uniformity
Encapsulation
Parameters
Dosator Volume
Attributes
Weight Variation
Capsule Potency
Content Uniformity
Dissolution
Interaction
Critical
Key
To be confirmed

# Continuous Quality Verification

# Continuous Quality Verification Key Concepts

- Science-based Approach to process validation
- Product and process understanding
  - Design Space
- Quality Risk Management
- Validate what is critical to product quality
- Continuous Improvement

# Continuous Quality Verification (ASTM E2537)

Pfizer

*"Continuous Quality Verification (CQV) is described as an approach to process validation where manufacturing process (or supporting utility system) performance is continuously monitored, evaluated and adjusted as necessary".*

"It is a science-based approach to verify that a process is capable and will consistently produce product meeting its pre-determined <u>critical</u> quality attributes (CQAs)."

With real time quality assurance (that CQV will provide), the desired quality attributes are ensured through continuous assessment during manufacture. Data from production batches can serve to validate the process and reflect the total system design concept, essentially supporting validation with each manufacturing batch"

ASTM E2537: Standard Guide for the Application of Continuous Quality Verification to Pharmaceutical and Biopharmaceutical Manufacturing

# CQV and Control Strategy

# Case Study Example: Translating Design Space into Control Strategy

| Dissolution | | | | | | |
|---|---|---|---|---|---|---|
| **Quality Attribute** | **Type** | **Process Parameters Impacting Quality Attributes** | **Type** | **Normal Operating Target and/or Range[1]** | **Design Space Range (PAR)** | **Control Strategy** |
| Disso-lution | CQA | API Particle Size Distribution | TBD | Particle Size Distribution TBD | Particle Size Distribution TBD | Release test - laser diffraction |
| | | Excipient Cold Water Solubility | TBD | TBD | A-B % cold water solubility | Release test by vendor |
| | | Initial Blender Speed | CPP | D rpm[2] | C-E rpm[2] | Set in recipe, monitored, alarmed, & trended in SCADA |
| | | Lubricant Feeder #1 Position | KPP | Fixed | Fixed | Fixed |
| | | Roller Compactor Force | CPP | G kN/cm (gap set at I mm, gap control on) | F-H kN/cm (gap set at I mm, gap control on) | Set in recipe, monitored, alarmed, & trended in SCADA |
| | | Roller Compactor Gap Width | KPP | Fixed at I mm (with gap control) | Fixed at I mm (with gap control) | Set in recipe, monitored, alarmed, & trended in SCADA |
| | | Mill Screen Size | KPP | J mm screen (fixed) | J mm screen (fixed) | Physically installed |
| | | Lunbricant Feeder #2 Position | KPP | Fixed | Fixed | Fixed |

KPP = Key Process Parameter
CPP = Critical Process Parameter
KQA = Key Quality Attribute
CQA = Critical Quality Attribute
TBD = To Be Determined
PAR = Proven Acceptable Range

# Benefits of Continuous Processing & Continuous Quality Verification

Pfizer

**Reduce**

- Scale-up
- Cycle-Time
- Processing/Testing Time
- Process Validation

**Improve**

- Capacity
- Manufacturing Facility Utilization
- Manufacturing Flexibility
- Process Understanding
- Trending
- Opportunities for Real-Time Release
- Regulatory Flexibility

# Real-Time Release

# Case Study Example: Real-Time Release

| Test | QC Method | On-line Strategy |
| --- | --- | --- |
| Identity | Visual | Visual |
| Identification | IR, HPLC | tbd (NIR?) |
| Potency | HPLC | Blend NIR and capsule weight |
| Content Uniformity | USP | Blend NIR |
| Dissolution Rate | HPLC (USP) | tbd |
| Water Content | Karl Fisher | tbd (NIR?) |
| Degradants | HPLC | tbd |

**tbd = to be determined**

# Knowledge Management

- We need to record and manage the knowledge we generate in the development of a process
  - This is done in the Process Understanding Plan (PUP)
- The PUP is used as substrate in the generation of the CTD
- Once filed and a product is launched the PUP is maintained by the manufacturing site and can be used to record further developments of the process.
- A feedback loop is put in place to allow learnings during development and post launch to be leveraged by the development and manufacturing functions

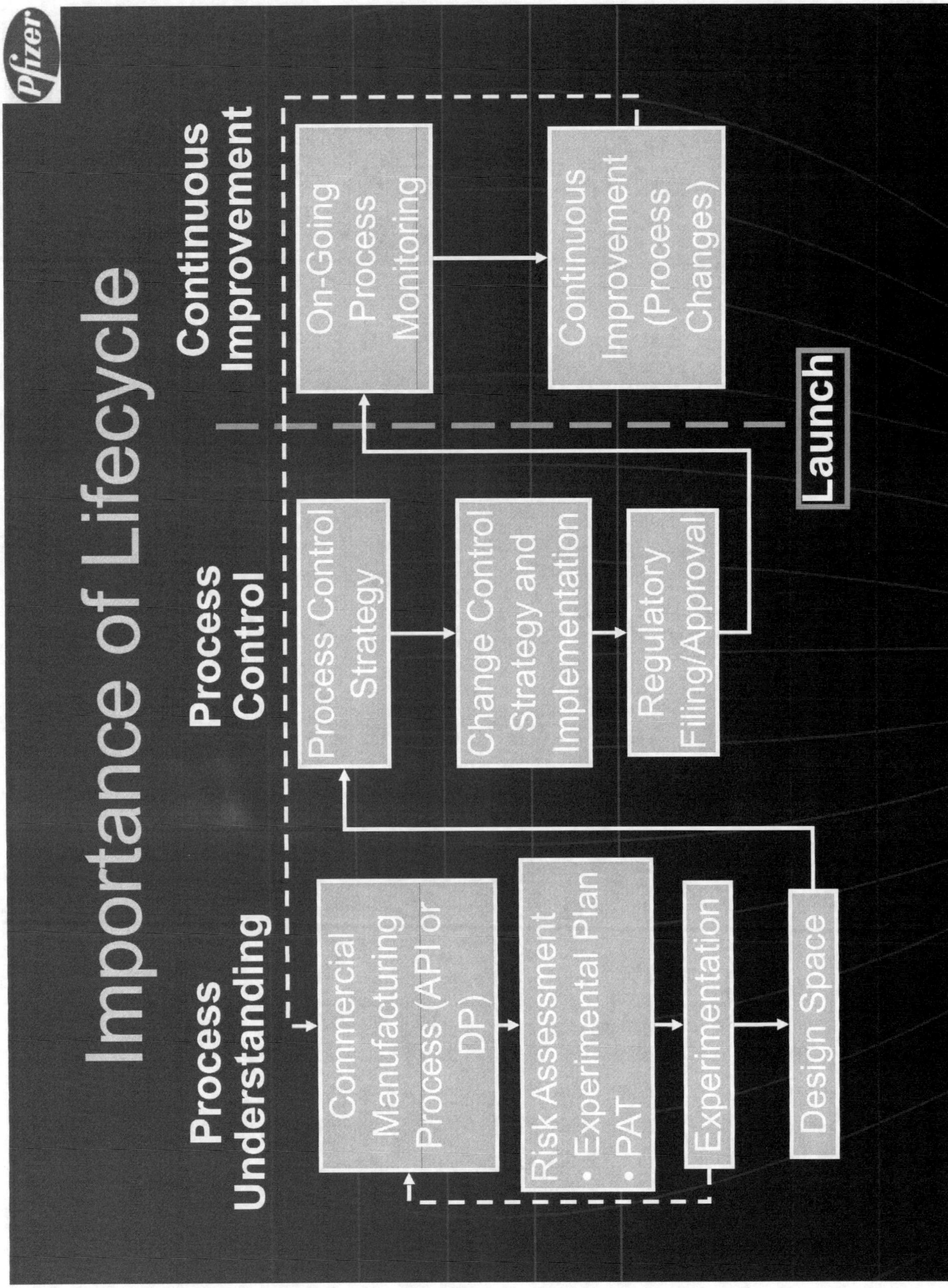
Importance of Lifecycle
Process Understanding
Commercial Manufacturing Process (API or DP)
Risk Assessment
• Experimental Plan
• PAT
Experimentation
Design Space
Process Control
Process Control Strategy
Change Control Strategy and Implementation
Regulatory Filing/Approval
Launch
Continuous Improvement
On-Going Process Monitoring
Continuous Improvement (Process Changes)

# Acknowledgments

- Kara Follmann
- Tom Garcia
- Al Pichieri
- Alex Chueh
- Mojgan Moshgbar
- Ke Hong
- Joseph Ceballos

# Multivariate Texture and Image Analysis for Solid Dosage Intermediate and Final Product Characterization and Downstream Performance Prediction

**Salvador Garcia-Munoz**, Process Modeling & Engineering Technology, Pfizer Global Research and Development, Groton, CT

Alan Carmody, Material Science and Oral Products Center of Emphasis, Pfizer Global Research and Development, Sandwich, United Kingdom

A framework using wavelet decomposition coupled with multivariate latent variable methods is proposed to quantitatively characterize the texture appearance of pharmaceutical materials and products. This framework allows having a non-biased measure of the visual characteristic of solid materials. The proposed technique has shown successful application in the steel sector , and is applied in a straight forward manner to grayscale images of the pharmaceutical products (or intermediate materials).

In summary, the technique uses wavelet decomposition on the grayscale images to reduce them to a vector of energies for multiple "frequencies" (called details in wavelet theory), such vectors are then compressed using Principal Components Analysis (for characterization) or used as regressors against a property of interest (e.g. in vitro dissolution, compression profile)

This method is illustrated with two case studies:

i) The analysis of an intermediate pharmaceutical product (wet granules from a high shear wet granulation step) to provide a quantitative envelope for the material, and also to predict its manufacturability at the compression step (after drying and milling).

ii) The characterization of a modified release film coated tablets, where the textural characteristics of the tablet are linked to the in vitro dissolution of the product.

Using a latent variable regression model, it was found that the appearance of the wet granules is able to predict 80% of the hardness compression profile for the material. In the case of the modified release tablets, the visual characteristics of the tablet alone can predict dissolution at 9hr with a 91% R2 accuracy. A model that considers the processing conditions in the film coater and the visual appearance is able to predict dissolution to a 97% R2.

See more of this Session: Qbd Across Drug Product Operations
See more of this Group/Topical: Topical I: Comprehensive Quality by Design in Pharmaceutical Development and Manufacture

# Modeling and Simulation in Support of Quality by Design

**Chau-Chyun Chen**, R&D, Aspen Technology, Inc., Burlington, MA

Quality by Design leverages a set of tools intended to ensure that a designed manufacturing process will consistently produce good quality product through the lifetime it is commercially manufactured. To ensure that such designs are robust through process scale-ups and other sources of variability in the commercial manufacturing process, manufacturers seek to understand and define the design space and then to further define and design the associated control spaces and control strategies. Fundamental to the tool set are advanced modeling tools. This includes both first principle modeling as well as empirical modeling.

This presentation reviews how modeling tools have been used in a variety of industries, in order to establish the design space and ultimately ensure robust commercial manufacturing processes.

See more of this Session: QbD Principles Applied in Other Industrial Sectors
See more of this Group/Topical: Topical I: Comprehensive Quality by Design in Pharmaceutical Development and Manufacture

# Model Based Fuel Cell Development

**Sitaram Ramaswamy**, United Technologies Corporation (UTC) Power, South Windsor, CT

To be added

See more of this Session: QbD Principles Applied in Other Industrial Sectors
See more of this Group/Topical: Topical I: Comprehensive Quality by Design in Pharmaceutical Development and Manufacture

# Model Base Design in the Fast Moving Consumer Goods

**Benjamin Weinstein**, Modeling and Simulation Department, Procter & Gamble, West Chester, OH

To be added

See more of this Session: QbD Principles Applied in Other Industrial Sectors
See more of this Group/Topical: Topical I: Comprehensive Quality by Design in Pharmaceutical Development and Manufacture

# DESIGN FOR SIX SIGMA IN HIGH VALUE MATERIALS DEVELOPMENT

George Liebermann
Senior Engineering Fellow/Black Belt

Xerox Research Centre of Canada
2660 Speakman Drive
Mississauga, Ontario, L5K 2L1, CANADA

## Abstract

*Design for Six Sigma (DfSS)* is an off-shot of the very successful *Six Sigma* methodology implemented by many major corporations to address opportunities in *improving existing processes and products* via a structured methodology for project management, problem solving and continuous optimization. Following the success of Six Sigma, a number of methodologies for development of *new products and processes* have been devised under the name of Design for Six Sigma. The overall principles of DfSS are discussed and the use of the methodology and its tool box will be detailed, in particular how it can be applied to high value specialty chemicals and polymers. Several examples will highlight the most useful DfSS tools that can be applied to projects in the chemical or related industries.

## INTRODUCTION

Xerox Research Centre of Canada (XRCC) has the global corporate mission to research and develop new materials for Xerox's imaging products such as laser printers and copiers, solid ink printers, as well as other specialized imaging devices. These materials can be described as high value specialty chemicals and polymers that are of electronic grade purity and have specific functional properties that make them useful in specific imaging technologies. Typical examples are photoactive materials, electron or hole transport molecules and specialty polymers used in photoreceptors; polymer resins with specific rheological and melt properties, pigments and charge control additives used in toners and low molecular weight polymers, rheology modifiers and dyes used in solid inks.

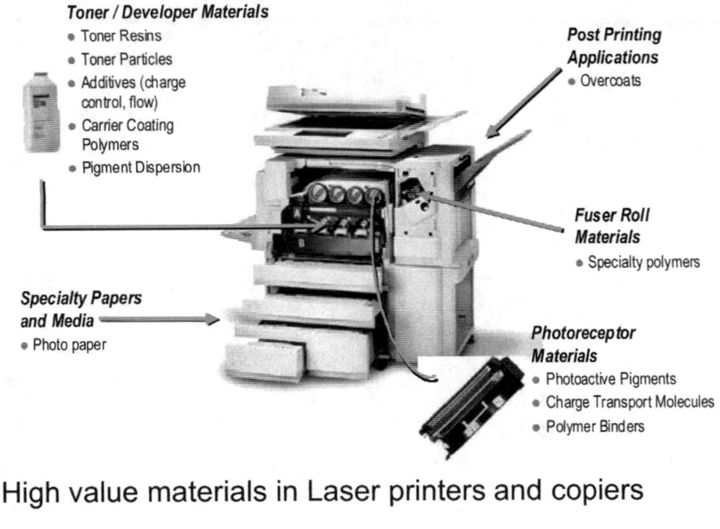


Figure 1. High value materials in Laser printers and copiers

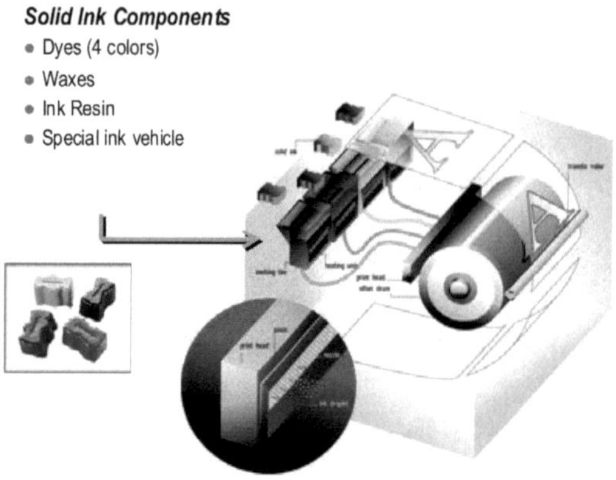


Figure 2. High value materials in Solid Ink printers

The advanced purity of the imaging materials and their functional requirements often results in challenges that can be in a certain way similar to the pharma chemicals: high purity and standard physico-chemical testing are not sufficient to determine if a particular molecule or polymer, or a specific lot will perform. A functional testing is normally undertaken, that can require the fabrication of a device or composite, and can be tested/evaluated on a range of testing apparatuses, bench surrogate imaging devices, breadboards, subassemblies, or a real imaging machine such as a laser printer.

The development of new materials for imaging applications poses complex challenges and our research centre has developed capabilities and competencies that have been critical in delivering many critical materials for commercialization in Xerox products since its inception in 1974. An illustration of these contributions in the last 10 years is presented in Figure 3.

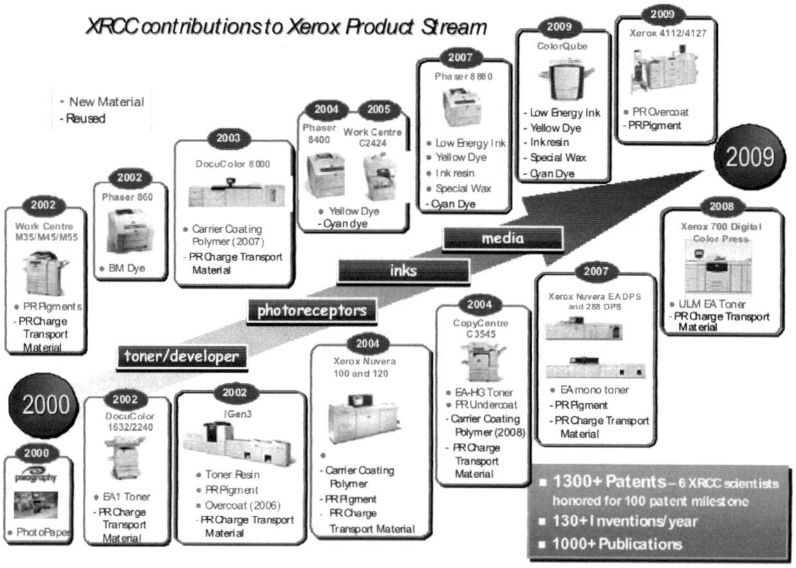


Figure 3. High Value materials in Xerox products

A significant change in the way Xerox is doing business was the Company-wide introduction of Lean Six Sigma methodology in 2003/2004, followed by the Design for Lean Six Sigma methodology in 2005/2006. A major impact in our research organization was the implementation of Design for Lean Six Sigma, the methodology targeting an improved, structured approach for the development of new products and processes.

## SIX SIGMA and DESIGN for SIX SIGMA (DfSS)

Six Sigma is a structured methodology which was adapted by major corporations as a framework for continuous improvement and problem solving. It uses a variety of tools and methods as it has evolved from Total Quality Management (TQM), procedures to reduce defects, various work process tools and statistics.

One of the most recognized characteristics of the Six Sigma methodology is the use of a 5 phase roadmap: DMAIC, which stands for the roadmap of 5 phases one will have to go through in a structured way for a particular project: Define, Measure, Analyze, Improve, Control.

Figure 4. Six Sigma/DMAIC

Integration of “Lean” manufacturing and work process practices with the statistics biased Six Sigma resulted in a methodology called Lean Six Sigma (LSS).

The success of Six Sigma methodology in problem solving and improvement of *existing processes/products* lead to a search and development of a similar methodology for *new processes and products* – the methodology was named Design for Six Sigma (DfSS)

While Six Sigma is well established with its DMAIC phases, there are several versions of DfLSS methodologies as far as phase numbers and names (roadmap), for example:

DMEDI – Define/Measure/Explore/Develop/Implement

DMADV - Define/Measure/Analyze/Design/Verify

IDOV – Identify/Design/Optimize/Validate

ICOV – Identify/Characterize/Optimize/Validate

Irrespective of the name and number of phases, all DfLSS methodologies are very similar and are using many tools already deployed by Six Sigma.

Expected benefits from using DfSS methodologies and tools in R&D include:

- early problems identification – solution when cost of the fix is low
- lower development cost
- robust product
- faster market entry

## DESIGN for SIX SIGMA (DfSS) in HIGH VALUE MATERIALS R&D

As DfSS is a general methodology and it provides a tool box with a large number of process and statistical tools, it is essential to adapt and customize its implementation to the specifics of the R&D environment while maintaining its key features. For a chemical or related industry these features can be translated as follows:

1. Determine *"Customer Requirements"* which in chemical R&D translates into specifications or tentative specifications

2. *"Process Thinking"*, i.e. understand the procedure, process that needs to be developed

3. *"Data Driven"*, decisions to be done based on data, use of statistics, target reduction of variations, use of statistical process control

4. *"Structured Approach"*, meaning running a project through well defined phases (roadmap), with checkpoints between phases

5. *"Tool Box"*, a relatively large number of process and statistical tools available out of which the relevant tools should be used according to the specifics of the project or task

Generally for DfSS use to be successful it is expected that management is championing it, that people are trained as Six Sigma Green Belts and Black Belts, and that a team approach is always used.

The implementation of DfSS at our materials research centre was initiated by training over 95% of the eligible population over a period of 3 years. A medium level training (Green Belt) and application of the DfSS methodology and tools to ongoing projects as part of the certification as Green Belt was essential to make DfSS pervasive through the organization. Just over 10% of the eligible population was trained at the higher level (Black Belt) for more in-depth statistical and leadership background.

As a result of this effort several DfSS tools have got a wide use in all our current projects and in mid 2009 we saw the first new material developed using DfSS tools in our labs on a project initiated in 2006 being commercialized in a Xerox printer.

The key elements found adapted by our chemists and chemical engineers in using DfSS methodology and the most useful tools are being summarized below.

The DfSS methodology version implemented involves the 4 phase IDOV roadmap:

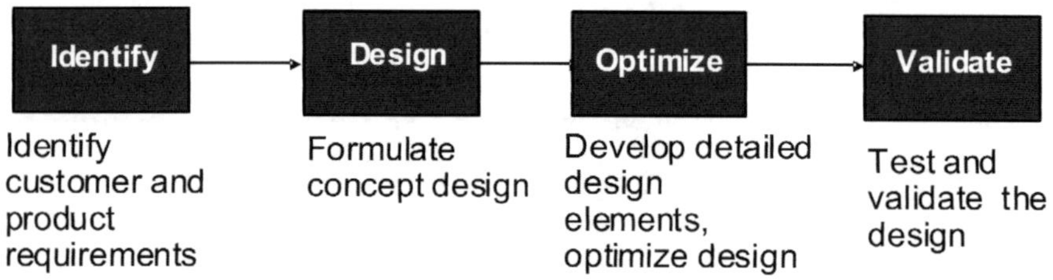


Figure 5. Design for Six Sigma/IDOV

The activities targeted in each phase are reasonably self-explanatory by the short description for each phase in Figure 5. It was found, however, that a high level of customization was required for each project, but the set of tools used was quite similar. Likewise, due to the fluidity of many research projects, in many occasions a full IDOV roadmap was not feasible and it became evident that in these cases the use of a set of tools without the formality of the IDOV roadmap can get the best results. However, the red-line of each project should largely follow the IDOV content.

### Phase 1. Identify customer and product requirements

The customer can be external or and internal to the company. Getting the customer requirements right involves first of all defining/identifying the framework of the project or study. There is a simple Six Sigma process tool for this called "Charter" which is a one page document that summarizes: **why, what, goal, how, when, who**

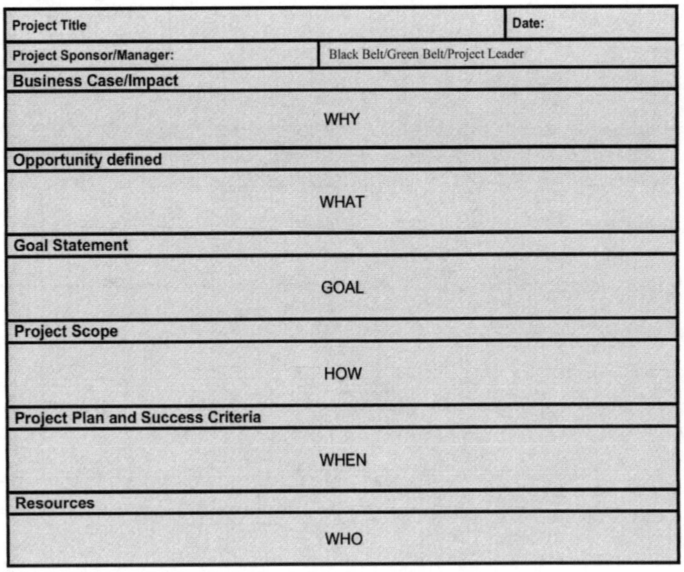


Figure 6. Typical DfSS Charter

The next step is to understand the internal/external customer requirements, their relative priority and translate them into what measurements will be required (analytical, characterization, functional) and what should be the target specifications. The tool recommended for this process described in Figure 7 is a "House of Quality".

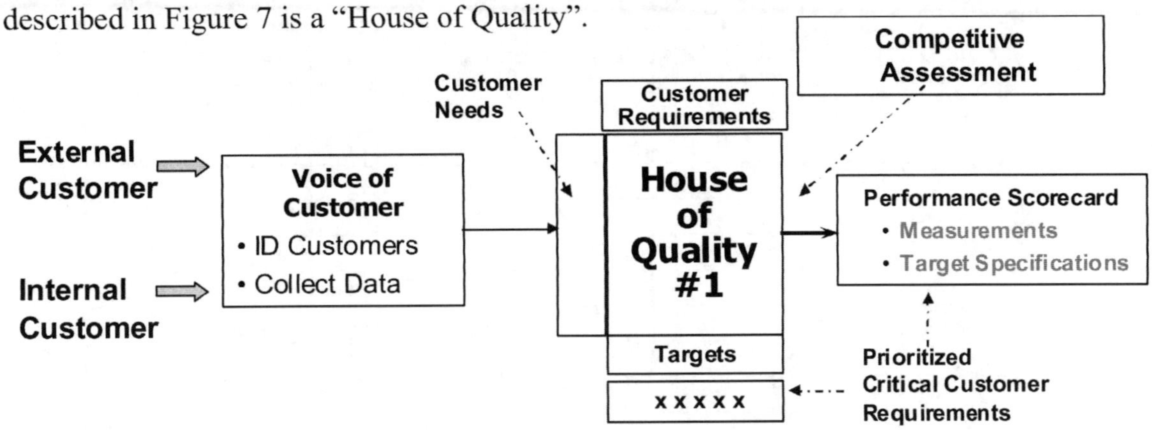


Figure 7. Measurements and Specifications

Having the right methods of measurements that are accurate, precise, stable, non-biased and with a good resolution is critical and a very important tool to determine this is a Measurement System Analysis (MSA). An MSA assesses how much variation is associated with the *measurement system* and compares it with the *process variation (and specifications).* The main tool is called "Gage R&R" assesses the percentage measurement variation (repeatability/same operator and reproducibility/different operators) represents of the variation of the process. Software packages like Minitab and JMP have standard MSA (Gage R&R) programs.

### Phase 2. Formulate concept design

The two main activities that recommended take place before any experimentation are the concept or solution selection and potential risk assessment.

The key tool for the concept/solution selection is the "Pugh Matrix". Several options are listed and evaluated against a baseline solution based on evaluation criteria and their weight. One or more concepts can be selected for detailed study.

1. List Requirements/Evaluation Criteria
2. Assign Weight to each Evaluation Criteria
3. List Solutions to be assessed (< 8-10)
4. Current process is the Baseline
5. Fill in Pugh Matrix
6. Rate Solutions vs baseline: +, 0, -, or ++, +, 0, -, --
7. Compute
8. Compare Totals (Pareto)
9. Select solution

**Pugh Matrix** - Which solution is the best

| | | Solutions | | | | |
|---|---|---|---|---|---|---|
| Evaluation Criteria (VOC/VOB) Quality, Cost, Delivery | Weight (Importance Ranking) (1-5) | Baseline (Current process) | Solution 1 | Solution 2 | Solution 3 | Solution 4 |
| | | | | | | |
| | | | | | | |
| | | | | | | |
| | | | | | | |
| | | | | | | |
| | | | | | | |
| | | | | | | |
| **Total** | | | | | | |

Figure 8. Pugh Matrix for concept/solution selection

For risk assessment of each of the proposed concepts or solutions there is a standard, but flexible tool named FMEA (Failure Modes and Effects Analysis)

| Proposed Solution/Change | Potential Failure Mode | Potential Failure Effects | SEVERITY | Potential Causes | OCCURANCE | Current Controls | DETECTION | RPN | Actions Recommended | Resp. |
|---|---|---|---|---|---|---|---|---|---|---|
| Description of change as per proposed solution | Negative Impact, Failure, or Error | Impact on Output Variables (identified risk items ) | | What could cause the problem | | Existing controls and procedures | | | To avoid/reduce occurrence and/or improve detection | |
| | | | | | | | | | | |
| | | | | | | | | | | |
| | | | | | | | | | | |
| | | | | | | | | | | |
| | | | | | | | | | | |
| | | | | | | | | | | |
| | | | | | | | | | | |

Figure 9. FMEA form

## Phase 3. Develop detailed design elements, optimize design

This phase is very specific for different projects and this is where most of the experimentation is carried out in the chemistry labs, benchscale chemical engineering or pilot plant. Most frequently used tool is design of experiments (DOE), that may be a screening DOE or a modeling or optimization DOE. Modeling and simulation of process steps, mixing, unit operations can be also big contributors to the success of the project. A simplified rendition of this phase is shown in figure 10. It goes without saying that most of the time, this is the longest and most demanding part of any project.

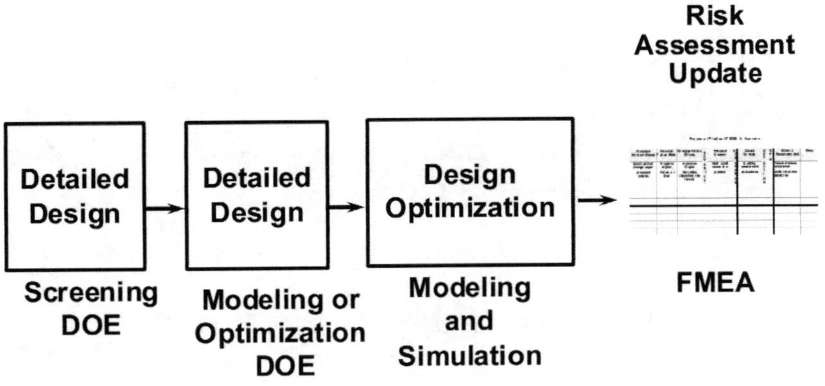


Figure 10. Detailed design and optimization

## Phase 4. Test and validate the design

In the final stage, demonstrating the design/solution at a larger scale, its reproducibility and validation against specification, safety and other requirements is usually done within company developed guidelines. At minimum the steps illustrated in Figure 11 are taken for each new process or product.

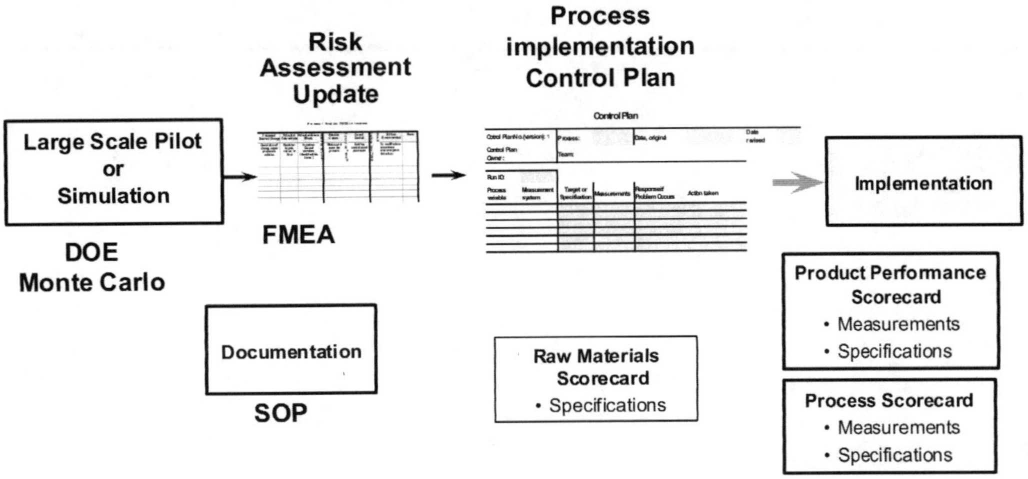


Figure 11. Validation

In summary, the main aspects of the Design for Six Sigma methodology used in a high value, electronic grade materials research and development activity were reviewed. The key message is that the chemical and related industries and chemical engineering are data driven, very well suited for applying Six Sigma methodologies. Applying the methodology and the tools creatively, where and when deemed to help, and in conjunction with sound technical judgment is critical for success.

# Experiences with Multivariate Quality Assurance in Dow

**Leo H. Chiang**, Core R&D, The Dow Chemical Company, Freeport, TX

This talk with present a variety of projects were multivariate latent variable methods were used to monitor and improve the quality of various products from Dow

See more of this Session: QbD Principles Applied in Other Industrial Sectors
See more of this Group/Topical: Topical I: Comprehensive Quality by Design in Pharmaceutical Development and Manufacture

# QbD for Biotherapeutics: Opportunities and Challenges -- An Industrial View Point

**Stefanie Pluschkell**, Gcmc, Pfizer Inc, Groton, CT

This paper will present an industrial viewpoint of the opportunities and challenges that have surfaced to date in the application of quality by design (QbD) principles in the development of biologics for human therapy. While a number of tools of product and process design already have a long tradition in the development of drugs derived by biotechnology, the development of the more recent ICH guidelines Q8, Q9 and Q10 has provided a regulatory framework for a comprehensive and prospective quality by design approach that fully integrates product and process development with the principles of quality risk management and quality systems. Quality by design can be applied throughout the lifecycle of a product. While QbD can begin as early as in the first stages of the engineering of the biotherapeutic itself and its formulation, extensive process design studies are typically not undertaken until the product reaches proof of concept in clinical trials. Case studies will be presented and some of the challenges of applying QbD to multiple unit operations and identifying the functional relationships between process input parameters and quality attributes of the drug will be highlighted. Different ways of developing and presenting a multidimensional manufacturing design space will be shown. Since the sources of heterogeneity of a biotherapeutic compound, such as a glycosylated protein, primarily originate in the drug substance manufacturing process, this presentation will focus on the production and purification of bulk formulated drug substance. Differences in the approaches that might be taken for a pre-approval versus a marketed product will also be illustrated. Well-characterized biologics, such as monoclonal antibodies, are good candidates for a comprehensive QbD approach, however, the stage of development, the type of anticipated post-approval changes, and a variety of other technical and business aspects will impact the decision of how thoroughly the design space will be explored prior to marketing approval. Both the advantages of increased process understanding and the outlook of reduced regulatory hurdles to continuous process improvement justify working through the challenges. Developing a process design space and control strategy that in the future may underwrite, for example, changes in equipment, process scale, or manufacturing site are often high on the priority list for manufacturers of biologics.

See more of this Session: 13 QbD in Bio Process Development and Operation
See more of this Group/Topical: Topical I: Comprehensive Quality by Design in Pharmaceutical Development and Manufacture

# A Systems Approach for Design of Intensified Bio-Pharmaceutical Processes

**Alicia Román Martínez**, Department of Chemical & Biochemical Engineering, Technical University of Denmark, Lyngby, Denmark

Rafiqul Gani, Department of Chemical & Biochemical Engineering, Technical University of Denmark, Lyngby, Denmark

John M. Woodley, Center for BioProcess Engineering,Department of Chemical & Biochemical Engineering, Technical University of Denmark, Lyngby, Denmark

The use of biocatalytic process technologies as a supplement to conventional chemical synthesis has gained particular attention in recent years since new options for sustainable and environmentally benign processes can be established. Biocatalysis offers numerous advantages for achieving end quality specifications, such as high purity, high-yield and high selectivity, and ambient conditions of operation. Despite the advantages that biocatalytic technologies offer, there are still many challenges for their sustainable implementation, such as the integration of the biocatalytic step with neighboring operations, which can be additional chemical reactions (one-pot synthesis) and/or separation operation (in-situ product removal, ISPR). Besides, the implementation of an industrial-scale bioprocess usually takes considerable time for development some of which can be reduced by using established process systems engineering techniques, such as model-based and computer aided synthesis/design techniques. These techniques need to be adopted for applications in bio-pharmaceutical process development.

In this work a systematic model-based generic methodology for design and development of intensified bioprocesses satisfying the product quality parameters as indicators for selection is presented. The methodology focuses on the (bio)reaction(s)/downstream processing stages of a bioprocess and considers ISPR and one-pot synthesis procedures as well as technical and economical aspects, to achieve the goal of intensified process development that will maximize the product yield, selectivity, productivity, biocatalyst stability and product purity as well as profitability, while trying to minimize energy requirements by employing fewer processing steps and higher yields in each step. This methodology consists of seven steps. Step 1: Problem definition, where the objectives for quality by design are established. Step 2: Collection of all known data for the system under consideration. Step 3: Development and validation of a generic model to be used for process-product analysis. Step 4: Generation of the intensified alternatives using synthesis/design algorithms. Step 5: Verification of the alternatives trough model-based simulation. Step 6: Optimization of the feasible alternatives to identify the optimal. Step 7: Validation of the optimal alternative for final selection.

Application of the developed model-based methodology will be described and highlighted for the synthesis of N-acetylneuraminic acid (Neu5Ac). Neu5Ac is an interesting pharmaceutical intermediate due to its anti-viral, anti-cancer and anti-inflammatory functions. The synthesis of Neu5Ac is a particularly interesting example since it is characterized by an unfavorable equilibrium, high amount of waste per kilogram of product and difficult downstream processing. Therefore, improved feasible alternatives of Neu5Ac synthesis have been generated, modeled and validated using this methodology.

See more of this Session: 13 QbD in Bio Process Development and Operation
See more of this Group/Topical: Topical I: Comprehensive Quality by Design in Pharmaceutical Development and Manufacture

# Upstream Process Characterization Utilizing Response Surface Experimental Design: Case Study for a GS-NS0 Monoclonal Antibody Production Process

Serena Fries Smith, Bioprocess Research and Development, Merck & Co. Inc, West Point, PA

Edith Senderak, Merck & Co. Inc, West Point, PA

Rachel Bareither, Merck & Co. Inc, Rahway, NJ

Melanie Tellers, Merck & Co. Inc, West Point, PA

Vanessa Parikh, Merck & Co. Inc, Rahway, NJ

T. Craig Seamans, Merck & Co. Inc, Rahway, NJ

Michel Chartrain, Merck & Co. Inc, Rahway, NJ

Process characterization is a systematic investigation into a production process and yields valuable information on performance and product quality attributes. The main goals of the characterization studies are to identify Key Operating Parameters (KOP) and Critical Process Parameters (CPP), define operating ranges for those parameters, and demonstrate process robustness. In addition to identifying acceptable operating ranges, setting recommended operating ranges within the acceptable limits is crucial to a robust process. Prior to beginning experimental studies aimed at defining the cell culture process, a Failure Mode Effects Analysis (FMEA) was completed based on previous process and equipment knowledge. Following the risk analysis, which prioritized the process parameters with the highest likeliness of failure, the characterization of the GS-NS0 fed-batch process was completed using a response surface experimental design in a 3-L stirred bioreactor scale-down model. The process parameters that were studied were temperature, pH, dissolved oxygen, inoculation density, feed timing, medium concentration, and feed concentration. These studies quantified the effects of the process parameters on cell growth, cell viability, antibody production, binding affinity, glycan distribution, and charge variant distribution, and identified temperature and pH as CPPs and inoculation density as a KOP. The other investigated parameters were determined to be non-key parameters. The characterization of this GS-NS0 fed-batch process provided necessary information to set recommended and acceptable operating ranges for the process parameters. It also highlighted process sensitivities and failure limits, which were useful during process scale-up and technology transfer to the manufacturing facility. Further, short-term excursion studies were performed to evaluate the effect of process control drifts beyond the acceptable operating range. The response surface experimental design worked well as a tool to characterize the fed-batch process. This design enabled the detailed mapping of operating space for the evaluated parameters and provided information necessary to identify the critical parameters for this process.

See more of this Session: 13 QbD in Bio Process Development and Operation
See more of this Group/Topical: Topical I: Comprehensive Quality by Design in Pharmaceutical Development and Manufacture

# Characterization of Monoclonal Antibody Primary Recovery and Purification Steps: Towards Process Validation

**Michael Iammarino**, MBV Process Development, Merck & Company, Inc., Rahway, NJ

Nihal Tugcu, MBV Process Development, Merck & Company, Inc., Rahway, NJ

Joseph Nti-Gyabaah, MBV Process Development, Merck & Company, Inc., Rahway, NJ

Martin Chandler, MBV Process Development, Merck & Company, Inc., Rahway, NJ

David Roush, MBV Process Development, Merck & Company, Inc., Rahway, NJ

This paper will focus on summarizing the approaches taken to characterize a monoclonal antibody downstream process including primary recovery and purification steps. Utilization of Failure Mode and Effect Analysis (FMEA) and experimental approaches to efficiently characterize primary recovery (centrifugation, depth filtration and ultrafiltration) and purification steps (chromatography) are presented. The important parameters (based on severity, occurrence and detection) were determined as a result of FMEA. In addition to probe experiments, design of experiment (DOE) methods were utilized where possible to increase the efficiency of experimentation and to make statistical analysis possible. As a result, the parameters were categorized into key, critical or non-key classes. The ranges identified were checked and process robustness was demonstrated via purification of off-center cell culture batches.

See more of this Session: 13 QbD in Bio Process Development and Operation
See more of this Group/Topical: Topical I: Comprehensive Quality by Design in Pharmaceutical Development and Manufacture

# Improving the Batch Cycle in Bioreactors: From Media Makeup to Cleaning

**Elaine Martin**, University of Newcastle upon Tyne., Newcastle upon Tyne, United Kingdom

Gary Montague, School of Chemical Engineering and Advanced Materials, University of Newcastle upon Tyne., Newcastle upon Tyne, United Kingdom

On a bioprocess plant where batch fermentation is the means of production, improving overall productivity requires consideration of all aspects of the batch operation: from charging media, through optimizing reaction, to completing reactor clean out. Achieving consistency in operation and demonstrating strict adherence to product quality specifications has always been of paramount importance to the pharmaceutical industry. There have been significant changes to the manner in which this achieved, with pressures now to move from a situation where the same recipe is followed to one where deeper process understanding informs the actions taken to achieve consistency. This new philosophy is termed Quality by Design (QbD) where fundamental knowledge on process behavior guides design decisions and operational policy. In adopting the QbD philosophy it is necessary to consider all the phases of batch operation and for each ensure that there is sufficient information available to operate in a responsive, robust and reliable manner. The increasing acceptance of on-line process analytical technologies (PAT) has significantly enhanced the information content available contributing to fundamental knowledge and industrial moves towards the QbD philosophy has ensured that new technology is accepted.

This paper considers the phases of batch operation and how process measurement, control and optimization strategies may be exploited for business benefit. The foundation for improvement is the utilization of the broad range of process measurements to provide the fundamental insight into process condition that is critical to the QbD concept. Typically maximizing information content involves process data fusion from spectroscopic type instruments and combing possible multiple signals from such devices with conventional process data. The paper presents strategies by which this may be achieved and demonstrates the additionality gained compared to using single data sources. Several approaches are contrasted for data compression and fusion adopting either linear or non-linear algorithms to extract pertinent information. How they cope with typical practical characteristics such as varying sample frequencies, batch length variations and process noise is presented. The importance of appropriate data pre-processing is demonstrated prior to the application of compression and fusion algorithms in order to make the analysis more tractable.

During operation, awareness of process state is essential information for the operators. One approach is to translate complex spectral signatures into concentrations measures using calibration modeling. An alternative is to use the available process data as a 'signature' of condition and apply data compression methods to indicate condition in an interpretable manner for the operators. Both strategies have their place as calibration models provide useful information to control systems to respond to process disturbance or demand changes whereas the fingerprinting approach is useful for proof of consistency in a comprehensive multivariate sense taking into account the numerous concentration changes that could potentially impact on behavior. This paper contrasts different strategies for calibration modeling and fingerprinting with emphasis on ease of use and interpretability. Again, linear and non-linear variants of the algorithms are considered.

Addressing the phases that make up the batch cycle involves consideration of the make-up of media and inoculation of a seed, the monitoring and control of batch progression and selection harvest time and finally the clean-out of the process plant prior to starting the next cycle.

In batch operation, achieving consistent initial conditions as far as possible involves attention to media preparation and seed inoculum transfer. PAT type measurements have found application in raw material acceptance testing but for fermentations making use of natural media components a degree of variability inevitably arises. The application of PAT to determine seed transfer time to inoculate the production vessel offers distinct advantages. Here moving from time based transfer where seed condition varies to a transfer based on seed condition where inoculum age varies offers greater consistency in production performance. PAT measurement provides an indication of the physiological state which is a more pertinent indicator of performance on transfer to the production vessel. Indeed, previous studies have indicated that consistency in seed is crucial to production vessel behavior as sub-standard seeds cannot be compensated for by subsequent operational changes. The use of PAT devices and NIR in particular for seed transfer and seed quality performance impact will be demonstrated in the presentation.

Following inoculation, the availability of new measurement devices now provides greater insight into state profiles giving vital information on accumulation of media components or where levels are falling to limiting concentrations. The improvement of all phases of batch operation involves ensuring reaction proceeds in robust and consistent manner and disturbances, whether from media effects or process deviations, are adequately compensated for. Whereas previously off-line analysis was the primary source of such data, on-line spectroscopic measurements mean greatly enhanced frequency and significantly reduced measurement delay and consequently they facilitate a move from open-loop or limited closed loop through operator intervention to one where closed loop control of broth compositions is feasible. Again, examples of the use of spectroscopic devices for state profile measurement and control will be provided. Success depends upon the availability of accurate calibration models and the methods of data fusion described previously are shown to be beneficial in obtaining more precise measurements.

Following reaction, vessel emptying and clean out is undertaken. Ensuring this aspect of the cycle is optimized is also crucial to overall plant performance. In some instances, the clean out phase is significantly longer than the reaction phase hence focus on improvement by minimizing cleaning times can have a significant impact on overall plant yield. In cleaning, tracking of process state involves monitoring the approach to cleanliness and determining when the plant achieves this condition. The traditional strategy has been to adopt an open loop approach and clean the process in the same manner from batch to batch. The consequence of using this strategy is that while the process plant is clean, it is likely that excessive time and energy have been used. The QbD concept applied to the cleaning operation would necessitate a move from open loop to closed loop operation where cleaning progress is monitored and cleaning terminated when cleanliness is achieved. The risk based aspect of the QbD philosophy requires that cleaning termination time determination involves specification of the risk of failure. The benefits come from moving from an open-loop strategy where risk is minimized but not quantified or awareness of risk may be limited to the closed loop strategy where risk is understood and balanced against other operational considerations such as time taken and cost. Strategies for monitoring the progression of cleaning and forecasting the end-point are presented.

The paper draws on results from a number of industrial case studies both in pilot plant and production facilities to show how, with appropriate measurements and methodologies to interpret the data, operational benefits may be gained. Emphasis is placed on the practicality of the instrumentation and the information it provides, addressing in particular the different challenges faces during process development and process manufacture.

See more of this Session: 13 QbD in Bio Process Development and Operation
See more of this Group/Topical: Topical I: Comprehensive Quality by Design in Pharmaceutical Development and Manufacture

# A Contemporary Perspective of Design Space Development and Maintenance

**James C. Spavins**, Global Chemistry, Manufacturing & Controls, Pfizer Inc, Groton, CT

Abstact: To be added

See more of this Session: Regulatory Aspects of QbD: Development and Maintenance of Design Space (Invited Presentations)
See more of this Group/Topical: Topical I: Comprehensive Quality by Design in Pharmaceutical Development and Manufacture

# The FDA - Novartis CRADA: A Risk-Based and Systems Approach for Quality by Design (QbD) Implementation for An Approved Pharmaceutical Product

**James Cheney**, Global PAT Group, Novartis Pharmaceuticals Corporation, East Hanover, NJ

Mansoor A. Khan, Cder/Ops/Otr/Dpqr, Food and Drug Administratin, Silver Spring, MD

Marc Goeller, Global PAT, Novartis, East Hanover, NJ

Chris Balducci, Global PAT Group, Novartis Pharmaceuticals Corporation, East Hanover, NJ

Lorenz Liesum, Global PAT Group, Novartis Pharmaceuticals Corporation, East Hanover, NJ

The principles of "Quality by Design" were applied to a commercial drug substance and drug product, in order to implement the FDA driven initiative "Pharmaceutical cGMPs for the 21st Century". The Project was also a Cooperative Research and Development Agreement (CRADA) with the U.S, Food and Drug Administration.

In the presentation we will show the application of ICH Q9 for Risk Management. The development of additional knowledge on the various steps of the processes through the use of Design of Experiments and the application of PAT technologies will be highlighted. The interaction with the FDA - Division of Product Quality Research will be discussed. In the end the interactions between Novartis with the review functions of FDA and EMEA on the supplements that were filed will be discussed.

See more of this Session: Regulatory Aspects of QbD: Development and Maintenance of Design Space (Invited Presentations)
See more of this Group/Topical: Topical I: Comprehensive Quality by Design in Pharmaceutical Development and Manufacture

# QbD - Design Space Management / Maintenance

**T. Kourti**, GlaxoSmithKline and McMaster University, Hamilton, ON, Canada

Design Space Management is very important in the product lifecycle as there are several issues to be addressed.

The first issue is of course the fact that the design space is expected to evolve after the initial submission as more data become available. Evolution of the Design Space may also mean an evolving control strategy; it is expected that more data and increasing process knowledge would result in better (and smarter) control of the process.

Other issues include change of scale, site transfer, change of suppliers for API and excipients, change in the covariance of the raw material characteristics within the same supplier.

Intentional changes, like opportunities to use soft sensors instead of PAT, or to expand the current PAT capabilities, should also be considered.

This work deals with solutions to problems that can be addressed under the framework of design space management. It demonstrates that a systematic methodology for maintenance will enhance confidence on the design space.

See more of this Session: Regulatory Aspects of QbD: Development and Maintenance of Design Space (Invited Presentations)
See more of this Group/Topical: Topical I: Comprehensive Quality by Design in Pharmaceutical Development and Manufacture

# A Strategy of Using Mechanistic Understanding for Continuous Quality Verification

San Kiang, Process Research and Development, Bristol-Myers Squibb Co., New Brunswick, NJ

**Shih-Ying Chang**, Research and Development, Bristol-Myers Squibb Company, New Brunswick, NJ

Quality by Design is a FDA initiative to mitigate quality risk of manufacturing through process understanding and control. In this presentation we will describe a strategy of combining risk assessment on key process steps, designing of monitoring tools to gain process insight, development of mechanistic model based on process physics, and verification through experimental design. Demonstrated by a case study of functional coating, we believe that such a strategy may be implemented. It is possible to acquire the benefits of fast scale up, streamlined technology transfer, and continuous quality verification with the support of regulatory flexibility.

See more of this Session: Regulatory Aspects of QbD: Development and Maintenance of Design Space (Invited Presentations)
See more of this Group/Topical: Topical I: Comprehensive Quality by Design in Pharmaceutical Development and Manufacture

# Role of Models for Implementation of Design Space: An FDA View

**Sharmista Chatterjee**, Onqda/Ops/Cder, FDA, Silver Spring, MD

Defining a multivariate design space is one of the key aspects of a Quality by Design (QbD) based drug development. Enhanced understanding gained from following a QbD approach could be leveraged to support control strategies such as Real Time Release Testing (RTRT). Mathematical models can be utilized throughout drug product development: beginning with risk assessment to identify potential parameters that may affect the finished product quality, mapping out and implementing a design space for process operation and finally to monitor/control the process during commercial manufacturing. The aim of this presentation is to illustrate, using relevant examples, various modeling approaches that can be used for the development and implementation of design space and for RTRT and to discuss some regulatory considerations for maintenance/management of these models throughout the product life cycle.

See more of this Session: Regulatory Aspects of QbD: Development and Maintenance of Design Space (Invited Presentations)
See more of this Group/Topical: Topical I: Comprehensive Quality by Design in Pharmaceutical Development and Manufacture

# Role of Real-Time Monitoring Tools and Chemometrics for Development and Maintenance of Pharmaceutical Design Space

**Huiquan Wu**, Division of Product Quality Research, HFD-940, FDA, Silver Spring, MD

Recent pharmaceutical quality initiatives (PAT, ICH Q8, Q8R1, Q9, and Q10) have highlighted the necessity and critical importance of using PAT tools and QbD principles for pharmaceutical development, process understanding, and design space development. Based on the screening results of risk analysis and design of experiments (DOE), real-time process monitoring tools such as vibration spectroscopic technologies could enable real time data acquisition for both critical process variables and key quality attributes. Chemometrics tools could help to extract critical and essential information from the process and product data. That information, once being integrated with the process and product engineering knowledge as well as control practice, could form the scientific basis of the design space concept. In this talk, several case studies based on FDA research activities will be presented to discuss the utilities of real-time monitoring tools and chemometrics for development and maintenance of pharmaceutical QbD design space.

See more of this Session: Regulatory Aspects of QbD: Development and Maintenance of Design Space (Invited Presentations)
See more of this Group/Topical: Topical I: Comprehensive Quality by Design in Pharmaceutical Development and Manufacture

# Drug Substance Scale and Equipment Assessments Linking a Design Space to a Manufacturing Facility: Case Studies

**Timothy J. Watson**, Chemical R&D, Pfizer, Inc, Groton, CT

to be added

See more of this Session: Qbd Topical Conference Plenary Session: Emerging Topics
See more of this Group/Topical: Topical I: Comprehensive Quality by Design in Pharmaceutical Development and Manufacture

# The Role of Automatic Process Control In QbD

**Richard D. Braatz**, Department of Chemical and Biomolecular Engineering, University of Illinois at Urbana-Champaign, Urbana, IL

This seminar will discuss the role of automatic process control in Quality by Design, with a primary emphasis on the crystallization of pharmaceuticals. Successful industrial applications of automatic process control strategies will be reviewed, as well as alternative control strategies that have been proposed for providing different tradeoffs between model fidelity, sensor accuracies, and robust performance. Collectively, the strategies will include feedback controllers based on solution concentration measurement via ATR-FTIR spectroscopy as well as particle measurements based on in-process laser backscattering.

See more of this Session: Qbd Topical Conference Plenary Session: Emerging Topics
See more of this Group/Topical: Topical I: Comprehensive Quality by Design in Pharmaceutical Development and Manufacture

# NIPTE-FDA Collaborative Case Study On Model-Based Design Space Development Across Scales and with Stability Considerations

**Gintaras V. Reklaitis**, Chemical Engineering, Purdue University, West Lafayette, IN

Hamid Arastoopour, Department of Chemical and Biological Engineering, Illinois Institute of Technology, Chicago, IL

Robin Bogner, School of Pharmacy, University of Connecticut, Storrs, CT

James K. Drennen III, Mylan School of Pharmacy, Duquesne University, Pittsburgh, PA

Lee E. Kirsch, Divsion of Pharmaceutics, The University of Iowa, Iowa City, IA

James Litster, Chemical Engineering, Purdue University, West Lafayette, IN

Eric J. Munson, Pharmaceutical Chemistry, University of Kansas, Lawrence, KS

Fernando Muzzio, Chemical and Biochemical Engineering, Rutgers University, Piscataway, NJ

Raj Suryanarayanan, Department of Pharmaceutics, University of Minnesota, Minneapolis, MN

Stephen W. Hoag, School of Pharmacy, University of Maryland, Baltimore, MD

Mansoor A. Khan, Cder/Ops/Otr/Dpqr, Food and Drug Administration, Silver Spring, MD

to be added

See more of this Session: Qbd Topical Conference Plenary Session: Emerging Topics
See more of this Group/Topical: Topical I: Comprehensive Quality by Design in Pharmaceutical Development and Manufacture

# QbD and Statistics: What a Chemical Engineer Needs to Know

**David Christopher**, Schering-Plough Research Inst, Kenilworth, NJ

to be added

See more of this Session: Qbd Topical Conference Plenary Session: Emerging Topics
See more of this Group/Topical: Topical I: Comprehensive Quality by Design in Pharmaceutical Development and Manufacture

# Model-Based Quality by Design Applied to An API Process

**Justin L. Burt**, Bristol-Myers Squibb, New Brunswick, NJ

James Bergum, Bristol-Myers Squibb, New Brunswick, NJ

Alan Braem, Bristol-Myers Squibb, New Brunswick, NJ

Antonio Ramirez, Bristol-Myers Squibb, New Brunswick, NJ

Lucius Rossano, Bristol-Myers Squibb, New Brunswick, NJ

Jose Tabora, Bristol-Myers Squibb, New Brunswick, NJ

Srinivas Tummala, Bristol-Myers Squibb, New Brunswick, NJ

In this case study, an API process was investigated within the Quality by Design framework. Employing a multi-staged experimental design as the basis of investigation, process chemistry models were developed using both mechanistic and empirical approaches. Statistical analysis of crystallization purging data facilitated the establishment of in-process limits for key impurities, and our chemical models were employed to generate design spaces based on these specifications. The predictive capacity of the models was explored both on the laboratory scale, and via comparison with production-scale batches. By employing a scientific and risk-based approach, an API design space was defined that ensures product quality to a high degree.

In broad terms, this API process can be considered as three steps: (1) chemical dehydration, (2) a sequence of reversible reactions that affords the API, and (3) reaction quench and isolation. Reaction modeling focused on the conversion of Intermediate Z, along with the formation of Impurities A and B arising from API degradation. The presence of Impurity C in the starting material introduces further complexity, since it is transformed to Impurities A, B, and D in proportions that vary as a function of API processing conditions. Concentrations of Intermediate Z and Impurity A must be controlled to set specifications in the isolated product, whereas Impurities B and D purge near-quantitatively.

Points that will be highlighted include: (1) comparison of the design spaces generated by the mechanistic and empirical models; (2) key insight gained via the mechanistic approach; (3) impact of upstream process challenges on API design space definition; (4) opportunities for combined empirical and mechanistic modeling approaches.

$$\text{Starting Material} \rightleftharpoons Z \rightleftharpoons \text{API} \longrightarrow A + B$$

$$\text{Impurity C} \longrightarrow A + B + D$$

See more of this Session: Model Based Qbd in Drug Substance
See more of this Group/Topical: Topical I: Comprehensive Quality by Design in Pharmaceutical Development and Manufacture

# Applications of Quality by Design Principles Utilizing Process Modeling and PAT in the Development of An API Process

Christopher L. Burcham, Chemical Product Research and Development; Process Definition, Eli Lilly and Company, Indianapolis, IN

Joeseph R. Martinelli, Chemical Process Research and Development; Process Definition, Eli Lilly and Company, Indianapolis, IN

Mark A. LaPack, Chemical Process Research and Development; Process Definition, Eli Lilly and Company, Indianapolis, IN

Process analytical technology (PAT) can be used as an online control tool to determine a process endpoint or to change a setpoint if installed as part of a feedback control loop. During the development of a process, it can be used to better elucidate the governing physics of a process, allowing for an understanding of the mechanism of the process. In the later example, this can lead to the development of process models and allows for the development of a process using the principle of "Quality by Design".

In this presentation, PAT is used in the development of a first principles process model. The use of the process model will be presented in the context of developing the synthetic process to make semagacestat, a compound currently in Phase III clinical trials for the treatment of Alzheimer's Disease.

The penultimate synthetic step in the semagacestat route involves the removal of a tert-butyloxycarbonyl (BOC) protecting group followed by the coupling of a carboxylic acid to the deprotected amine. During the deprotection, carbon dioxide (CO2) is evolved. In the subsequent coupling reaction, residual CO2 reacts with the deprotected amine, forming a urea dimer, if the CO2 has not been sufficiently purged. Control of the urea dimer impurity is accomplished by the thorough removal of CO2 prior to the initiation of the coupling reaction. A mass transfer model for the partitioning of CO2 in the reaction matrix and the reactor headspace was derived to predict the concentration of CO2 in the reaction matrix, as it not measurable at the low levels needed for suitable control of the impurity. The model requires a mass transfer coefficient and the Henry's law partitioning coefficient as inputs. These parameters were determined using online mid-IR, to measure CO2 in solution (at levels above that needed for control) and mass spectrometry to measure CO2 in the headspace. The model was used for optimize the processing parameters needed to purge residual CO2 prior to the onset of the coupling reaction. The model allowed for an understanding of the effect of scale-up to commercial scale on the purge process parameters, eliminating the risk of failure with scale-up. The model was verified at production scale, by monitoring the concentration of CO2 in solution by ATR IR and in the headspace by GC mass spec, resulting in excellent agreement between the measured results and expectations. The subsequent product exhibited no measurable amount of the urea impurity.

See more of this Session: Model Based Qbd in Drug Substance
See more of this Group/Topical: Topical I: Comprehensive Quality by Design in Pharmaceutical Development and Manufacture

# Enhancing Process Understanding around Impurities Formation in API Intermediate Synthesis Via Kinetic Modeling & CFD

**Samrat Mukherjee**, Process R&D, Abbott Laboratories, North Chicago, IL

Shailendra V. Bordawekar, Process R & D, Global Pharmaceutical Research and Development, Abbott Laboratories, North Chicago, IL

Nandkishor K. Nere, GPRD Process Engineering, Abbott Laboratories, North Chicago, IL

An API intermediate synthesis involving Mitsunobu coupling reaction was investigated in-depth to understand variation in impurity levels with scale. A detailed kinetic model was developed in DynoChem to enhance the understanding around formation of two major impurities. This model was used to guide the process parameter range finding activity. The precursor for one of the impurities is a reaction byproduct that co-crystallizes with another reaction byproduct, thereby introducing scale-dependent parameters such as mixing in the overall scheme. CFD was used to post-analyze the reaction data to throw more light into the aforementioned dynamics.

See more of this Session: Model Based Qbd in Drug Substance
See more of this Group/Topical: Topical I: Comprehensive Quality by Design in Pharmaceutical Development and Manufacture

# An Integrated High-Throughput Experimentation and Semi-Empirical Modeling Approach to Building Process Knowledge in Pharmaceutical Process Development

**Alexander J. Marchut**, Process Research and Development, Bristol Myers Squibb, New Brunswick, NJ

Jun Qiu, Process Research and Development, Bristol Myers Squibb, New Brunswick, NJ

Olav K. Lyngberg, Process Research and Development, Bristol Myers Squibb, New Brunswick, NJ

A. Erik Rubin, Process Research and Development, Bristol Myers Squibb, New Brunswick, NJ

We demonstrate the power of collaboratively integrating predictive modeling and high-throughput experimentation to deliver data needed for research and scale up in a pharmaceutical process research and development setting. Since semi-empirical models need empirical data as input and high-throughput experiments can be very effectively designed when they are informed by a model prediction, these approaches lead naturally to the following integrated workflow. A high-throughput experiment is done as the first step, a model is developed based on this data as a second step, the space of process options is explored using the model as a third step, and the most promising model results are examined in targeted high-throughput experiments as a final step. This workflow can be used to speed up development timelines and improve process understanding by reducing the number of experiments performed while at the same time providing a wealth of data about the process. A combination of high-throughput experiments and semi-empirical models provides the best option for this task as the possibilities for process options can be too many to fully study with even high-throughput experimentation, ab initio models can be too computationally intensive and not sufficiently accurate, and although the processing options can be fully studied rapidly with semi-empirical models, they require experimental measurements as input and should be experimentally verified. As an example, we will focus on high-throughput solubility measurements and solubility modeling with the NRTL-SAC (Non Random Two Liquid Segment Activity Coefficient) solubility model. We have extensively and routinely applied this workflow in building knowledge around solubility of new Active Pharmaceutical Ingredients (API's) and intermediates in the synthesis of API's. The impact of this collaboration in terms of time savings will be presented as well as an example of a typical result.

See more of this Session: Model Based Qbd in Drug Substance
See more of this Group/Topical: Topical I: Comprehensive Quality by Design in Pharmaceutical Development and Manufacture

# Numerical Simulation of Mixing and Drying of Pharmaceutical Particles in a Bubbling Fluidized Bed Reactor

**Jungkee Jang**, Chemical and Biological Engineering, Illinois Institute of Technology, Chicago, IL

Cezar A. Rosa, Chemical and Biological Engineering, Illinois Institute of Technology, Chicago, IL

Hamid Arastoopour, Department of Chemical and Biological Engineering, Illinois Institute of Technology, Chicago, IL

The bubbling fluidized bed flow regime is characterized by high heat and mass transfer rate and relatively short drying times. In this project, the gas-solid mixing and drying of pharmaceutical particles in a bubbling fluidized bed reactor was simulated using the commercially available FLUENT 6.3 Computational Fluid Dynamics (CFD) code. The simulations are based on the two phase model and on the Eulerian-Eulerian approach. The mathematical model used is composed of the continuity, momentum and energy equations. Additionally, the classic drying curve model is used which includes pre-warming period, constant-rate period, and falling rate period. An effective thermal conductivity is used in the energy balance. The expressions of thermal conductivity, particle density, and drying rate were added in a form of an executable code by means of a user defined function (UDF). Moisture transfer in each phase was simulated using a user defined scalar transport equation (UDS). In order to determine the optimum particle mixing, numerical simulations were performed at different particle diameters, bed heights, inlet velocities and inlet velocity distributions, respectively. Moreover, to better understand the effect and impact of gas distribution on gas-solid flow patterns, 3-dimensional simulations were carried out using a perforated gas distributor. Simulation results are compared with experimental data to validate the potential of our model and to refine it for scale up purposes.

See more of this Session: Model Based Qbd in Drug Substance
See more of this Group/Topical: Topical I: Comprehensive Quality by Design in Pharmaceutical Development and Manufacture

# Design Space, Models and Model Uncertainty

**Constantinos C. Pantelides**, Process Systems Enterprise Ltd., London, United Kingdom

Nilay Shah, Centre for Process Systems Engineering, Imperial College London, London, United Kingdom

Claire S. Adjiman, Centre for Process Systems Engineering, London

The concept of Design Space plays a central role in the current thinking on the regulation of pharmaceutical production processes. Most such processes involve large numbers of external disturbances and/or potential control manipulations. This results in multidimensional Design Spaces, whose effective determination and exploration require the use of process models. A range of different categories of models is currently being used, including empirical correlations, statistical non-parametric models, data-based models and first-principles models.

This paper focuses on first-principles models, of the kind that are finding increasing use in pharmaceutical applications. We consider the relation between the Design Space and the concept of "process flexibility" which has been studied extensively in the process systems engineering literature since the early 1980s. We are particularly interested in the extent to which existing flexibility analysis techniques can be applied to the determination of the Design Space, both in principle and in practice, given the transient and multistage nature of most pharmaceutical processes.

Design Spaces are supposed to provide assurances that the process will deliver products of the right quality provided the operation is maintained remain within a given envelope. However, like the result of any other model-based computation, the reliability of any Design Space determined on the basis of a model depends on the accuracy of the model itself. In other words, the Design Space is not a well delineated region, but a probabilistic one, with each point in the multivariable space of operation of the process being characterized only by a probability of belonging to the Design Space. We consider techniques for computing probabilistic Design Spaces based on quantitative information on model uncertainty derived during the formal validation of such models against experimental data sets.

See more of this Session: Model Based Qbd in Drug Substance
See more of this Group/Topical: Topical I: Comprehensive Quality by Design in Pharmaceutical Development and Manufacture

# Correlation of Material Properties and Process Parameters to Pharmaceutical Tablet Mechanical Characteristics Using a Quai-Continuum Model

**Athanas A. Koynov**, Rutgers University, Piscataway, NJ

Alberto Cuitino, Mechanical Engineering, Rutgers University, Piscataway, NJ

The compaction of granular materials is a process utilized by many industries from the ceramical and metallurgical to the chemical and pharmaceutical ones. In fact, the currently most prevalent device for drug delivery - the solid tablet is produced through the process of powder compaction. A clear understanding of the dependence of the mechanical properties of the solid product on both the characteristics of the starting materials and the system dynamics during the process itself is crucial for the development of a Design Space, in accordance with the Quality by Design initiative. For this purpose, a quasicontinuum approach has been implemented to simulate the compaction of confined heterogeneous granular systems consisting of particles of different sizes as well as made up from different materials. Quasicontinuum models rely on a finite element framework, in which the displacement of most constrained particles is computed based on the displacements of the nodes of an adaptive mesh. The method has been augmented to consider both cohesive forces and time-dependent powder particle bonding. This allows for the prediction of the mechanical characteristics of the compacted tablets.

See more of this Session: Model Based Qbd in Drug Product
See more of this Group/Topical: Topical I: Comprehensive Quality by Design in Pharmaceutical Development and Manufacture

# Flowchart Methodology Utilizing Validated Process Models as the Foundation for a Quality-by-Design (QbD)-Based Approach for Pharmaceutical Unit Operations: Example Case Studies

**Lisa J. Graham**, Bend Research, Inc., Bend, OR

Daniel E. Dobry, Bend Research, Inc., Bend, OR

Ronald A. Beyerinck, Bend Research, Inc., Bend, OR

A model-based process development methodology is described which uses a Quality by Design (QbD) approach to establish a manufacturing design space. A flowchart methodology is described that illustrates how to rationally step through process development, using fundamental and engineering models and experimental tools in order to define the relationship between process inputs (input variables and process parameters) and critical-to-quality attributes. A key to this methodology is defining the physical situation and establishing the key control volume that must be understood and recreated from batch-to-batch and upon transfer and scale-up. In addition, the process understanding and experimentally-validated models within this methodology form the basis of a PAT development strategy (including the process control strategy) upon commercialization. This QbD approach can be applied to a number of pharmaceutical unit operations. Three case studies are reviewed - evaporative processes, melt-spray congeal, and liquid mixing operations.

Spray drying to form solid dispersions will be used as a case study to illustrate the QbD process development methodology for evaporative processes. In this process, the control of drying rate and atomization are critical to product performance, stability and manufacturability. The keys to this development approach are to 1) characterize the physicochemical properties of the formulation, 2) define the thermodynamic operating space utilizing a mass and energy balance, and 3) characterize the key kinetic components of the process such atomization and drying rates using experiments and models. Similar logic can be applied to other evaporative processes such as tablet coating and fluid bed coating or granulation.

For a melt-spray-congeal (MSC) process, the control of microsphere size is imperative for attaining a specified and reproducible sustained-release profile. In the case of MSC microspheres created by atomizing a melt stream using a spinning disk and congealing them in a cool gas stream the process variables determining microsphere size are: 1) disk speed, 2) disk diameter, 3) melt feed rate and 4) the state of the melt stream. The state of the melt stream, namely its composition and temperature, determine its viscosity and surface tension and these in combination with disk speed, disk diameter and melt feed rate are the control variables of importance. By determining the combinations of process variables that reach the intended microsphere size one can establish process conditions that result in the desired product.

A liquid mixing process scale-up approach will be described that utilizes computational fluid dynamics and a universal mixer experimental tool. The keys to the approach are to 1) define the physical situation, 2) create models (e.g. CFD) of each appropriate reactor scale/geometry, 3) correlate small scale experiments to model outputs to define key scaling parameters and validate the models 4) utilize a validated correlation between experimental data and model outputs to define conditions for process transfer/scale-up.

See more of this Session: Model Based Qbd in Drug Product
See more of this Group/Topical: Topical I: Comprehensive Quality by Design in Pharmaceutical Development and Manufacture

# Model Based Approach to Determine Optimum Operating Conditions in High Shear Granulation

Defne Kayrak-Talay, School of Chemical Engineering, Purdue University, West Lafayette, IN

James D. Litster, School of Chemical Engineering; and Department of Industrial and Physical Pharmacy, Purdue University, West Lafayette, IN

James K. Drennen III, Mylan School of Pharmacy, Duquesne University, Pittsburgh, PA

This work is a part of FDA sponsored project that is carried out by Members of the National Institute for Pharmaceutical Technology and Education (NIPTE). The aim of the project is to develop Quality by Design (QbD) guidance elements on process design space, scale-up, and process validation for three unit operations: mixing, granulation, and drying; and to develope a framework for optimizing design space specification across scales with consideration to stability. The specific aim of this work is to implement scientific improvements to high shear wet granulation to determine the best operation window on a laboratory scale granulation system for a pharmaceutical formulation. The key dimensionless groups and regime maps developed from these dimensionless groups in the last decade are used to determine the regime of operation which will provide desired granule characteristics. Dimensionless groups, namely dimensionless spray flux ( a), dimensionless penetration time ( p), maximum liquid saturation (Smax) incorporate both formulation properties and process parameters. The formulation characteristics including particle size, true density, tap density, penetration time of binder into powder surface; and the process characteristics including powder surface velocity, binder drop size, circulation time in the granulator, flow rate of the liquid binder are determined, initially. Gabapentin is used as the active pharmaceutical ingredient and water is used as the liquid binder. The granulation formulation also consists of hydroxylpropylcellulose (HPC). This work consists of two parts. In the first part, laboratory scale experiments are designed and performed based on regime maps to determine good operating conditions as well as to show extreme operating conditions in Diosna (6 l) high shear granulator. In the second part, another granulator system (Gral) is investigated by using the same formulation. By keeping the dimensionless groups constant, the knowledge on Diosna granulator is transferred to Gral granulator. The process characteristics of the new system are explored to combine them with formulation properties in order to get the same product characteristics as it is obtained in Diosna.

See more of this Session: Model Based Qbd in Drug Product
See more of this Group/Topical: Topical I: Comprehensive Quality by Design in Pharmaceutical Development and Manufacture

# Modeling Density and Temperature Distributions in Powder Die Compacts

Jerry R. Klinzing, Materials Science and Engineering, Drexel University, Philadelphia, PA

Antonios Zavaliangos, Materials Science and Engineering, Drexel University, Philadelphia, PA

John Cunningham, Johnson & Johnson Pharmaceutical Research & Development, Spring House, PA

Tracey Mascaro, Johnson & Johnson Pharmaceutical Research & Development, Spring House, PA

Denita Winstead, Johnson & Johnson Pharmaceutical Research & Development, Spring House, PA

Tableting is one of the most important unit operations in the pharmaceutical industry because more than 90% of drugs are delivered in solid dosage forms. The physical and mechanical properties of tablets such as strength, density, and temperature are directly related to this process. The ability to predict through finite element (FE) modeling the physical and mechanical properties is essential towards maintaining a desired design space. For example, FE predictions of density distributions and temperature, which can alter dissolution rates and the polymorphic state of the active ingredient, respectively, could result in a change outside the design space. Therefore FE simulations are an important aspect of establishing a design space for support of quality by design for pharmaceutical processing. Previous FE modeling attempts were conducted in 2D (axisymmetry) but this current research explores the full 3D nature of the problem with the model material Avicel PH102.

The fully coupled thermomechanical FE software package ABAQUS provided a means to implement the Drucker-Prager Cap model which describes the evolution of the yield surface with varying relative densities. In addition, Young's Modulus and Poisson's ratio were varied with respect to RD to describe the elastic properties of the powder during unloading. A single compaction, unloading, and ejection step were analyzed for a capsule shaped tablet. Density variation and the temperature history of every point within the tablet is predicted. Furthermore, microcomputed tomography density measurements and infrared camera temperature measurements of post compacted tablets from a fully instrumented compaction simulator were used to validate the model.

The FE simulation agrees with the experimental force-displacement profile of the punches throughout compaction. The density predictions of the model are further validated with microcomputed tomography measured relative density values as shown through several internal cross sections. In addition, the model is capable of predicting surface temperatures of the post ejected tablets as supported by infrared camera temperature measurements.

The model is capable of predicting the full 3D nature of density and temperature distributions and further validates the implementation of the Drucker-Prager Cap model for pharmaceutical tableting. Moreover, the ability of FE to predict such properties presents the opportunity to further optimize tablet design, tool design, lettering placement, sticking, and overall process development.

See more of this Session: Model Based Qbd in Drug Product
See more of this Group/Topical: Topical I: Comprehensive Quality by Design in Pharmaceutical Development and Manufacture

# Film Coating End-Point Determination for Colored Immediate Release Tablets Using Multivariate Image Analysis

**Salvador Garcia-Munoz**, Process Modeling & Engineering Technology, Pfizer Global Research and Development, Groton, CT

Daniel Gierer, Material Sciences and Oral Products Center of Emphasis, Pfizer Global Research and Development, Groton, CT

Quality by design is an exercise that relies on quantitative metrics of the quality of a product. For immediate release (IR) film coated products, the uniformity of the coat is traditionally assessed by visual inspection. It is desired to substitute such assessment with a quantitative one.

In this work, Multivariate image analysis (MIA) was used to assess film coated product. The objective is to determine the end-point for cosmetic coating of the tablets, and calculate the distribution of coating across individual tablets as the process progresses; using digital RGB images as inputs. Knowing the distribution of the coating will allow the user to establish reasonable limits of coating material to be applied accounting for the mixing and spray inefficiencies in the coater.

This method has been successfully applied to snack food , and relies on the correlation of appearance with the property of interest. In the case of coated tablets there is a correlation between the weight increase in the tablets and their color. This correlation is broken when the coating is uniform, and the application of more coating material on the tablet does not result in a change in color (this point is referred to as the cosmetic end-point).

MIA is performed on digital images taken off-line, of a representative number of tablets sampled at different time points during the coating run; the percentage increase in weight is determined for each group of tablets. The images and the estimated weight increases are used in the MIA method to calibrate a predictive model to estimate the average coating weight for a given set of pixels in the image.

The average weight increase is calculated by applying this prediction to an image of multiple tablets. The coating distributions are calculated by subdividing the images into area units approximately equal to that of a single tablet. The prediction is done for each sub-image, and the set of predictions is used to produce a distribution.

Two lots of 100mg tablets from two different scales were studied to assess the off-line application of the method. As expected, the coating distributions on tablets produced at smaller scale exhibit narrower distributions that those at large scale. The estimated average amount of coating required for cosmetic purposes is the same in both cases, however differences in mixing (which result in different distributions) dictate a different cosmetic-end point for each one.

An initial assessment to carry out this analysis in-situ was also made using a digital video camera inside a laboratory scale coater. An adaptive PCA model is proposed as a method to monitor the changes in the appearance of the tablets. Experimental results are shown to illustrate the use of the method in successfully determining coating-end point for a 175mg round tablet.

[1] H. Yu, J. F. MacGregor, G. Haarsma, and W. Bourg, "Digital Imaging for Online Monitoring and Control of Industrial Snack Food Processes," Ind. Eng. Chem. Res., vol. 42, pp. 3036-3044, 2003.

See more of this Session: Model Based Qbd in Drug Product
See more of this Group/Topical: Topical I: Comprehensive Quality by Design in Pharmaceutical Development

and Manufacture

# The Effect of Tablet Shape On Film Coating Uniformity Using the Discrete Element Method

**William R. Ketterhagen**, Pfizer Global Research and Development, Pfizer Inc., Groton, CT

Film coating of pharmaceutical tablets is an important process that is conducted to help improve the tablet appearance, the stability of the active drug, taste masking, or, in the case of controlled release tablets, modify the release profile. Regardless of the rationale for applying the coating, the uniformity of the coating is an important attribute. For immediate release coatings where coatings may be applied primarily for aesthetic purposes, poor uniformity leads to reduced process efficiency and increases run times to ensure tablets with the least coating meet quality criteria while other tablets are over-coated. More serious consequences occur for functional membrane coatings that control the release of the active drug. In this case, variable coating uniformity may lead to large variability in the drug release rates.

In this work, the discrete element method (DEM) is used to computationally model the dynamics of tablet motion in a film coating pan. The number and duration of appearances in a fictitious spray zone are used to predict the coating uniformity under those conditions. A key variable explored in this work is the shape of the tablets, where different tablet shapes are approximated using the "glued spheres" technique. The results show that tablet shape does affect the typical tablet orientation in the spray zone, and thus, the intra-tablet coating uniformity. Finally, the effect of various process conditions such as pan rotation speed and pan loading are examined. Results show that pan speed affects the intra-tablet uniformity while pan fill level seems to have a negligible effect. These results demonstrate the usefulness of modeling in guiding drug product development decisions such as shape selection and process operating conditions.

See more of this Session: Model Based Qbd in Drug Product
See more of this Group/Topical: Topical I: Comprehensive Quality by Design in Pharmaceutical Development and Manufacture

# Controllability Analysis and Identification of Optimal Control-Loop Pairings in a Multiple-Input Multiple-Output Granulation Process

Rohit Ramachandran, Department of Chemical Engineering, Massachusetts Institute of Technology, Cambridge, MA

Charles D. Immanuel, Centre for Process Systems Engineering, Department of Chemical Engineering, Imperial College London, London, United Kingdom

Frantisek Stepanek, Department of Chemical Engineering, Institute of Chemical Technology, Prague, Prague 6, Czech Republic

Granulation is complex process whereby fine particles agglomerate to form larger granules, due to adhesive forces brought about by the liquid binder. Many of the existing granulation operations are in continuous mode and given the pharmaceutical industries' recent initiative to shift from batch to continuous, granulation in continuous operation assumes more importance [1]. In a continuous granulation process, feed material is continuously introduced into the granulator alongside the interplay of the various granule mechanisms such as nucleation, aggregation, consolidation and breakage. The granulator is fitted with several spray nozzles through which the liquid binder is introduced into the granule bed. The granules formed are then dried and classified based on product specification(s). Granules that do not conform to these specification(s) are recycled and reprocessed. Prior to the actual design and implementation of the controller for the granulation process, it is important to ascertain how well the process can be controlled and what factors may hinder the control-loop performance. It is also imperative that appropriate process inputs and outputs are selected for feedback control and that they are paired correctly as incorrect pairings may limit and hinder control-loop performance. A process is said to be controllable if there exists a controller that in principle, be able to achieve a certain output state via certain admissible input changes [2]. Controllability is an intrinsic process property and is determined by the process design such as sensor locations, recycle loops, internal couplings and sizing of equipment. Controllability conditions in granulation are also made more challenging given the presence of internal variables [3].

The aim of the present work is firstly to systematically ascertain the controllability of the granulation process by examining the effect of control handles such as binder and feed flowrates on controlled variables such as average size, distribution width, moisture content and bulk density. Secondly, to identify suitable control-loop pairings to facilitate a better understanding of the design and operation of the granulation process. Results are presented on a continuous and compartmentalized population balance model of a granulator, which depicted a realistic model of a pilot-plant granulation process. The model considered was based on granulation mechanisms that have been previously derived from first principles and validated at the batch-scale for different operating conditions and granulation systems [4,5]. Model simulations elegantly captured the nonlinear effect of nozzle positions on the granule output. As a direct consequence of the compartmentalization, the controllability of the granulation process could be examined, taking into account all the important process inputs and outputs, providing valuable insight into the control and operation of the granulation process. Results show that the average diameter, moisture content and bulk density are realistic choices of controlled variables given the previously identified manipulated variables at the operators' disposal. The size distribution width was not sensitive to any of the manipulated variables and deemed uncontrollable which also corroborates the experimental findings of Glaser et al. (2009) [6]. This is an important finding as size distribution is an important variable that needs to be tightly regulated. For instance, in the pharmaceutical industry, variations in granule size distribution (GSD) could lead to undesirable variations in tablet weight. As a result, this paves the way for alternative manipulative variables that have shown to influence the GSD as identified in the experimental work of Ramachandran et al. (2008) [7]. In this work, by means of step tests and a relative gain array analysis, several possible control-loop pairings were identified and within them, an optimal control-loop pairing was selected. Such a study on a

MIMO granulation process lends valuable insight and knowledge which can potentially be applied to an actual granulation process.

References

1. C. Vervaet and J. P. Remon, "Continuous granulation in the pharmaceutical industry", Chemical Engineering Science, 60, 3949-3957, 2005.

2. R. A. Eek and O. H. Bosgra, "Controllability of particulate processes in relation to sensor characteristics, Powder Technology, 108, 137-146, 2000.

3. D. Semino and W. H. Ray, "Control of systems described by population balance equations - i. controllability analysis", Chemical Engineering Science, 50, 1805-1824, 1995.

4. J. Poon, R. Ramachandran, C. Sanders, T. Glaser, C. D. Immanuel, F. J. Doyle III, J. D. Litster, F. Stepanek, F. Y. Wang and I. T. Cameron, "Experimental validation studies on a multi-dimensional and multi-scale population balance model of batch granulation", Chemical Engineering Science, 64, 775-786, 2009.

5. R. Ramachandran, C. D. Immanuel, F. Stepanek, J. D. Litster and F. J. Doyle III, "A mechanistic model for granule breakage in population balances of granulation: theoretical kernel development and experimental validation", Chemical Engineering Research and Design, 87, 598-614, 2009.

6. T. Glaser, C. F. W. Sanders, F. Y. Wang, I. T. Cameron, J. D. Litster, J. Poon, R. Ramachandran, C. D. Immanuel and F. J. Doyle III, "Model predictive control of continuous drum granulation", Journal of Process Control, 19, 615-622, 2009.

7. R. Ramachandran, J. Poon, C. Sanders, T. Glaser, C. D. Immanuel, F. J. Doyle III, J. D. Litster, F. Stepanek, F. Y. Wang and I. T. Cameron, "Experimental studies on distributions of granule size, binder content and porosity in batch drum granulation: Inferences on process modelling requirements and process sensitivities", Powder Technology, 188, 89-101, 2008.

See more of this Session: Automatic Control and Qbd
See more of this Group/Topical: Topical I: Comprehensive Quality by Design in Pharmaceutical Development and Manufacture

# Expanding the Acceptance Region for Incoming Materials Via the Usage of a Mid-Course Correction Controller for Wet Granulation

**Salvador Garcia-Munoz**, Process Modeling & Engineering Technology, Pfizer Global Research and Development, Groton, CT

Stephanie Brecht, Analytical R & D, Pfizer Global Research and Development, Groton, CT

Howard W. Ward II, Analytical R & D, Pfizer Global Research and Development, Groton, CT

Multivariate Latent Variable Methods are combined with advanced non-linear programming techniques to construct a feed-forward controller for the wet granulation operation. Four types of controllers are contrasted: i)feed-forward using only raw material properties, ii) feed-forward doing mid course corrections using power consumption and temperature measurements from the process, iii) same as ii with the addition of a NIR signal, iv) feed-forward doing mid course corrections using exclusively the NIR signal. Results show that the best performance is obtained when the controller uses the three trajectories available (power, temperature and NIR). These results are then used to illustrate how the acceptance criteria for the raw materials will be affected by the control strategy.

See more of this Session: Automatic Control and Qbd
See more of this Group/Topical: Topical I: Comprehensive Quality by Design in Pharmaceutical Development and Manufacture

# Application of ICAS-PAT On Design of Process Monitoring and Control System for a Batch Cooling Crystallization Process through Hybrid Multiscale Model-Based Analysis

**Ravendra Singh**, CAPEC, Department of Chemical and Biochemical Engineering, Technical University of Denmark, 2800 Kgs. Lyngby, Denmark

Noor Asma Fazli Bin Abdul Samad, CAPEC, Department of Chemical and Biochemical Engineering, Technical University of Denmark, 2800 Kgs. Lyngby, Denmark

Gürkan Sin, CAPEC, Department of Chemical and Biochemical Engineering, Technical University of Denmark, 2800 Kgs. Lyngby, Denmark

Krist V. Gernaey, BioEng, Department of Chemical and Biochemical Engineering, Technical University of Denmark, 2800 Kgs. Lyngby, Denmark

Rafiqul Gani*, CAPEC, Department of Chemical and Biochemical Engineering, Technical University of Denmark, 2800 Kgs. Lyngby, Denmark

A well designed process monitoring & control systems insures that the critical process variables are measured and maintained within the design space and therefore predefined end product quality can be achieved precisely and consistently. Design of process monitoring and control system involves the identification of the critical quality parameters, selection of economical & reliable data measurement methods/tools and implementation with the efficient control systems. Singh et al., (2009) proposed the use of a systematic model and knowledge based methodology to design/validate the PAT systems (process monitoring & analysis systems) and implemented this methodology into a systematic computer aided framework to develop a software (ICAS-PAT) to make the PAT design procedure simpler, better and faster. The software ICAS-PAT is now extended to incorporate the detailed control algorithm, needed to design/validate the product quality through implementation of efficient and advanced control systems. In the control algorithm, options for steady state and dynamic analysis are implemented. The steady state analysis is developed in order to identify candidate operational scenario and analyzed using steady state sensitivity analysis. This analysis is then transferred to corresponding dynamic model to assess controllability performances using steady state and dynamic relative gain array and interaction between controlled and manipulated variable. The extended version of ICAS-PAT is therefore now capable to provide the product recipe that can be directly use in the manufacturing floor to produce the product with desired quality. The main feature of new software is the design of PAT systems. However, some additional features are also developed to provide the options to open and analyze the solved examples, to identify the different applications of the monitoring techniques/tools, to generate the possible alternatives of the PAT systems, to generate a report in MS-Word and to generate closed/open loop process flow-diagrams (indicating also the monitoring and control schemes).

The objective of this presentation is two-fold, first to highlight the extension of the framework and the corresponding software (ICAS-PAT) and then to demonstrate their application through a batch cooling crystallization process. The extended ICAS-PAT is integrated with two supporting tools: an extended knowledge base and a generic model library needed to provide the necessary knowledge/data for design of process monitoring and control systems. The extended knowledge base consists of two sections. The first section of the knowledge base summarizes the necessary process knowledge (type of processes, corresponding process points, process variables and actuators) while the second section of the knowledge base consist of the knowledge on measurement methods and tools (type of variables, available monitoring techniques and tools with specifications such as accuracy, precision, operating range, response time, resolution, cost etc.). The model library contains a set of mathematical models for different types of unit processes, sensors and controllers.

The application of ICAS-PAT and the new framework will be illustrated through a batch crystallization process, a most common unit operation used when isolating the final product. Usually the quality specification for crystalline substances is the desired crystal size distribution (CSD). The intriguing problem here is to determine how to monitor & control the crystallization operation during a batch run such that a desired CSD is obtained at the end of the batch. As needed for design & validate the process monitoring & control system, a generic hybrid multiscale batch crystallization process model has been developed and included in model library.

Reference: Singh, R., Gernaey, K. V., Gani, R., (2009). Model-based computer-aided framework for design of process monitoring and analysis systems. Computers & Chemical Engineering, 33, 22-42

PAT: Process Analytical Technology

*Corresponding author (rag@kt.dtu.dk)

See more of this Session: Automatic Control and Qbd
See more of this Group/Topical: Topical I: Comprehensive Quality by Design in Pharmaceutical Development and Manufacture

# Feed-Forward Process Control Strategy for Pharmaceutical Tablet Development

**Koji Muteki**, Pfizer Inc, Groton, CT

Vidya Swaminathan, Pfizer Inc, Groton, CT

Sonja Sekulic, Pfizer Inc, Groton, CT

A feed-forward process control strategy which can compensate for the effect of raw material lot-to-lot variability on final tablet attributes in pharmaceutical tablet development is presented. The approach involves building PLS (partial least squares) models that combine raw material attributes and tablet process parameters to relate these to final tablet attributes. Prior to a manufacturing batch, the resulting models are used in an optimization framework to find optimal process conditions which can satisfy desired final tablet attributes while taking into account the attributes of specific raw material lots used in the batch. The effectiveness of the methodologies is illustrated through an industrial tablet manufacturing case study.

See more of this Session: Automatic Control and Qbd
See more of this Group/Topical: Topical I: Comprehensive Quality by Design in Pharmaceutical Development and Manufacture

ENGINEERING RESEARCH CENTER FOR
STRUCTURED ORGANIC PARTICULATE SYSTEMS
RUTGERS UNIVERSITY
PURDUE UNIVERSITY
NEW JERSEY INSTITUTE OF TECHNOLOGY
UNIVERSITY OF PUERTO RICO AT MAYAGÜEZ

# Automation of a Fluid Bed Dryer utilizing Multi-Variate Near Infrared Spectra and Model Predictive Control

Ariel O. Vargas Colon, Dr. Carlos Velázquez
University of Puerto Rico
Mayaguez Campus
Department of Chemical Engineering

# Agenda

- Introduction
- NIR Sensor
- System Integration
- System Model
- MPC
- Results
- Questions

# What is Near Infra Red Spectroscopy (NIRS) ?

- Is a technique that measures the absorption of radiation of a sample

- Advantages of NIR
  - Identify and quantify
  - High Speed process
  - Low cost
  - Non- destructive

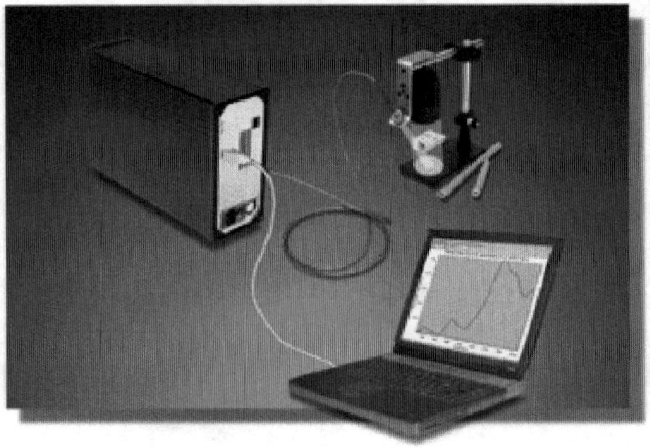

# Online NIR sensor

- Multiple spectrum are needed to verify sensor usefulness
- Chemometrics is needed to predict moisture
- Model needs to adjust for powder movement

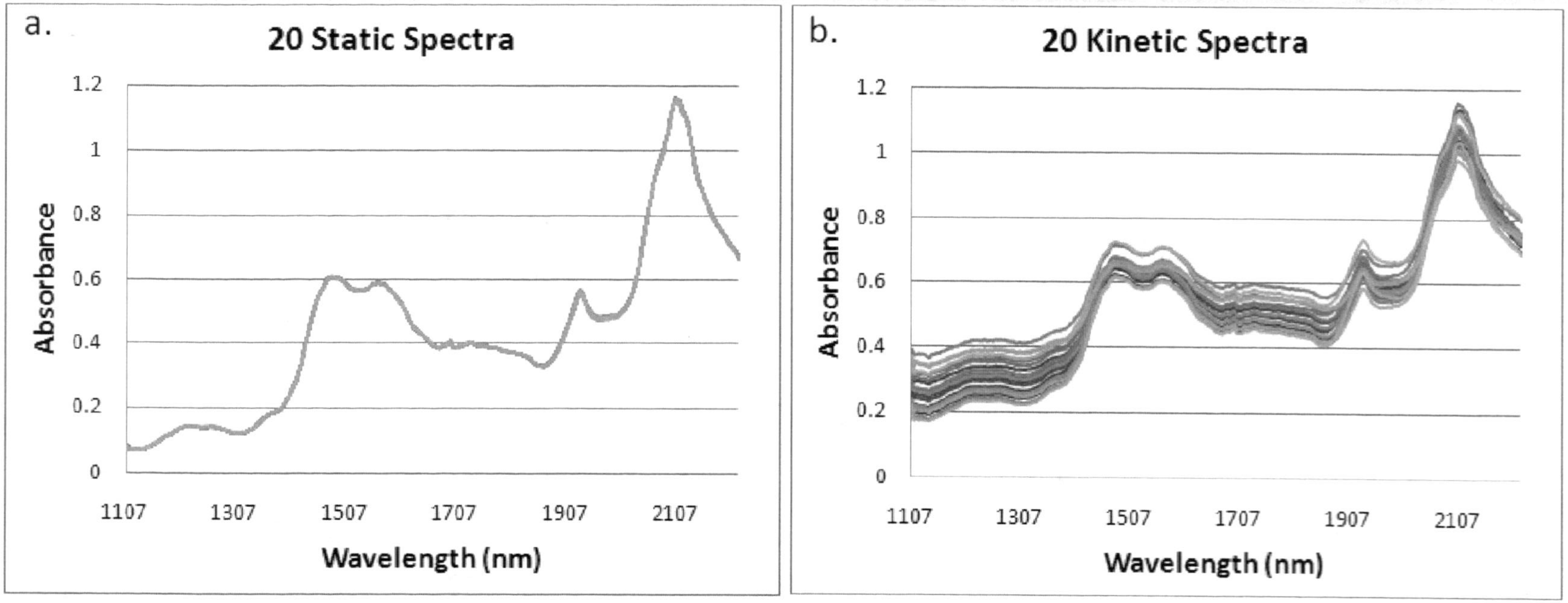

# PLS model generated

Three models where generated.

- Predicts moisture content in powders
- Predicts the error in the moisture prediction
- Predicts the error of the error predicted

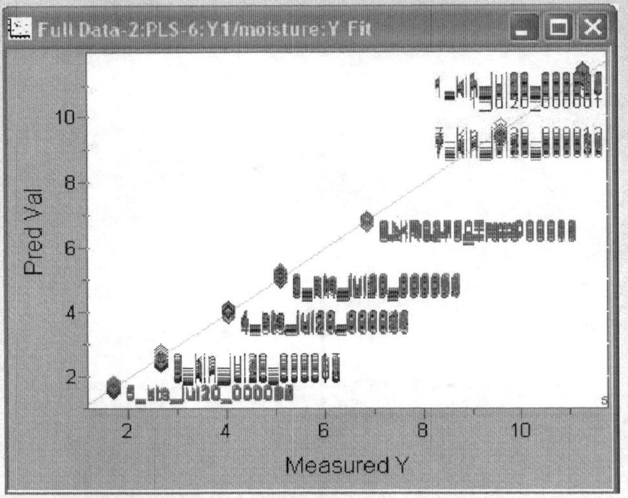


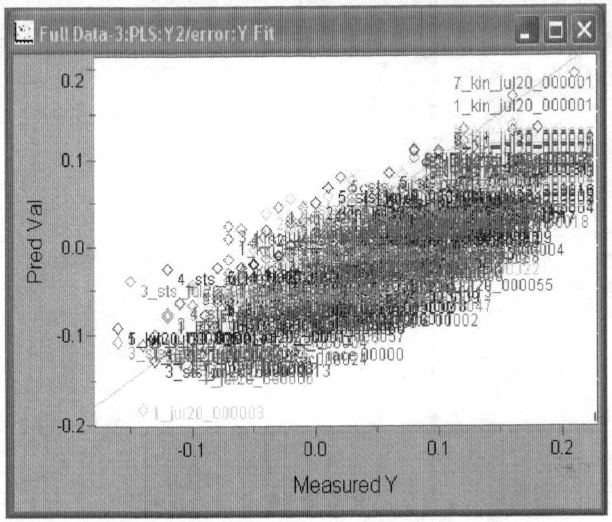


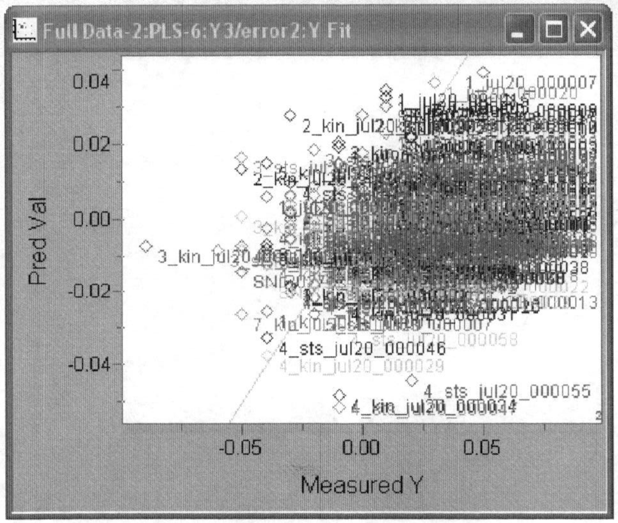

# NIR prediction

- The mathematical treatment improves NIR moisture Prediction
- The application of the method can be done every 3 seconds.

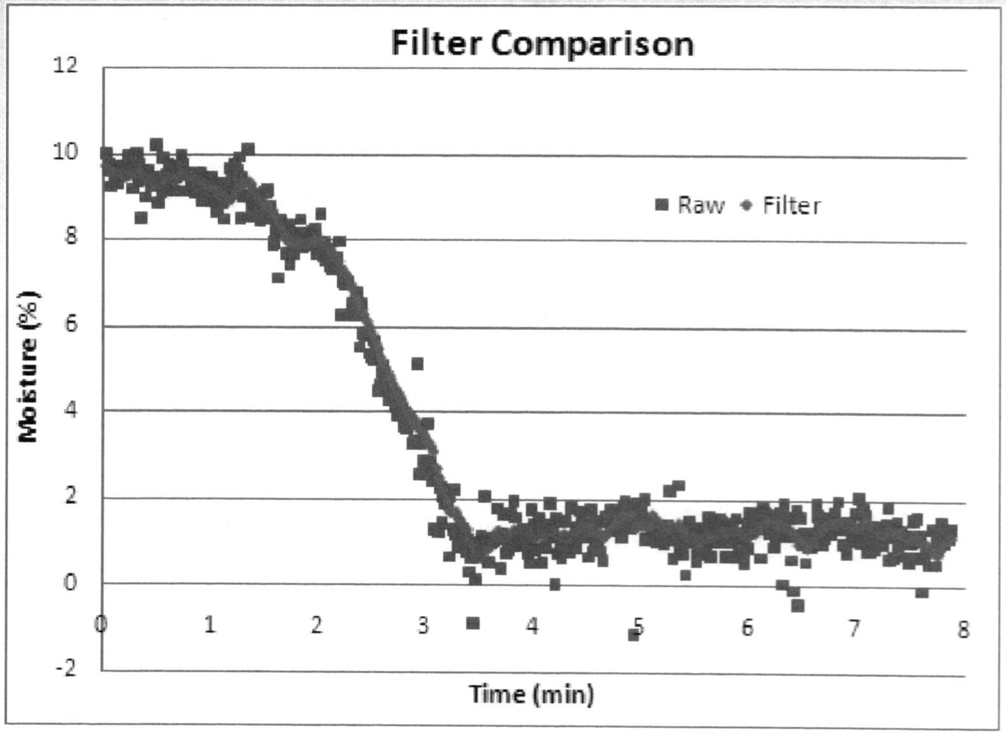

# Model validation

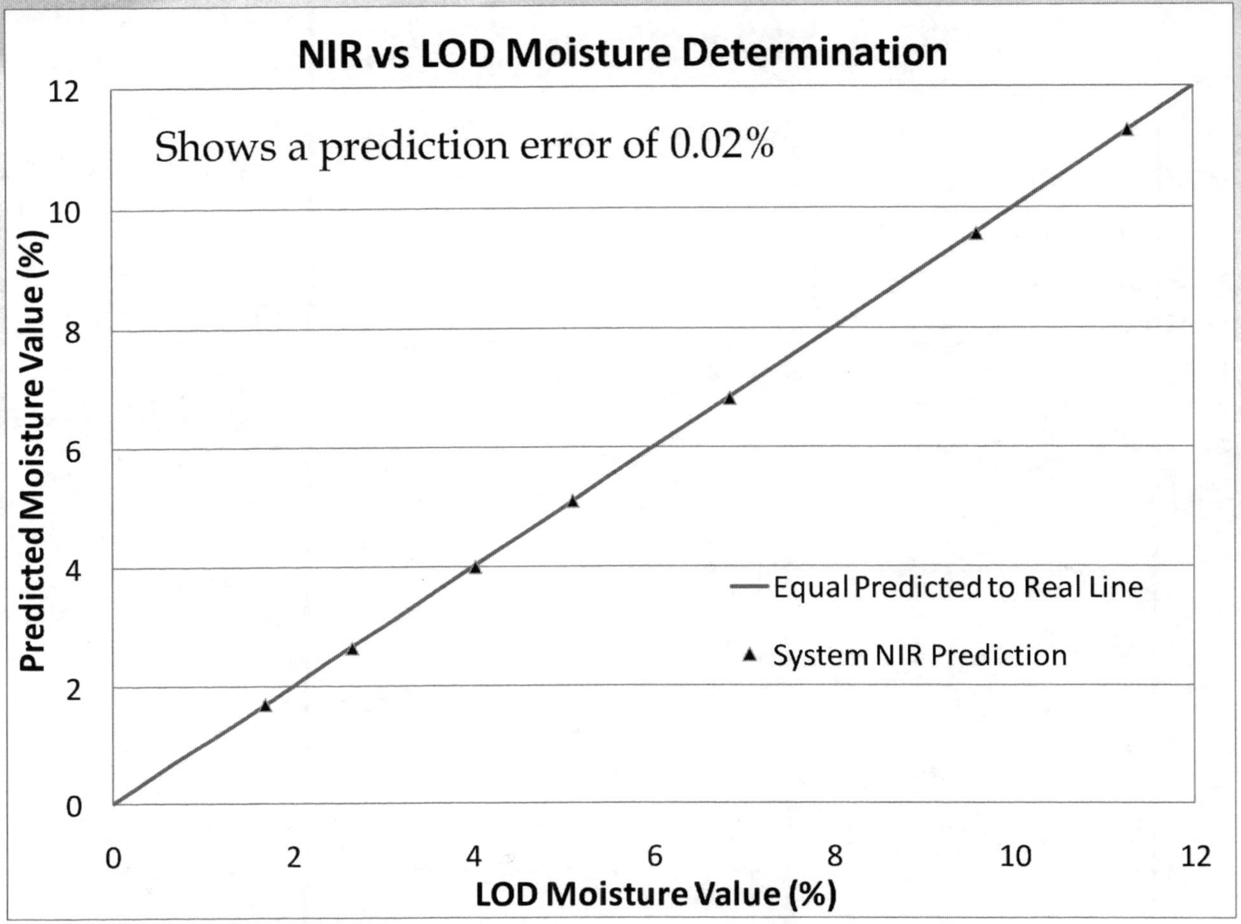

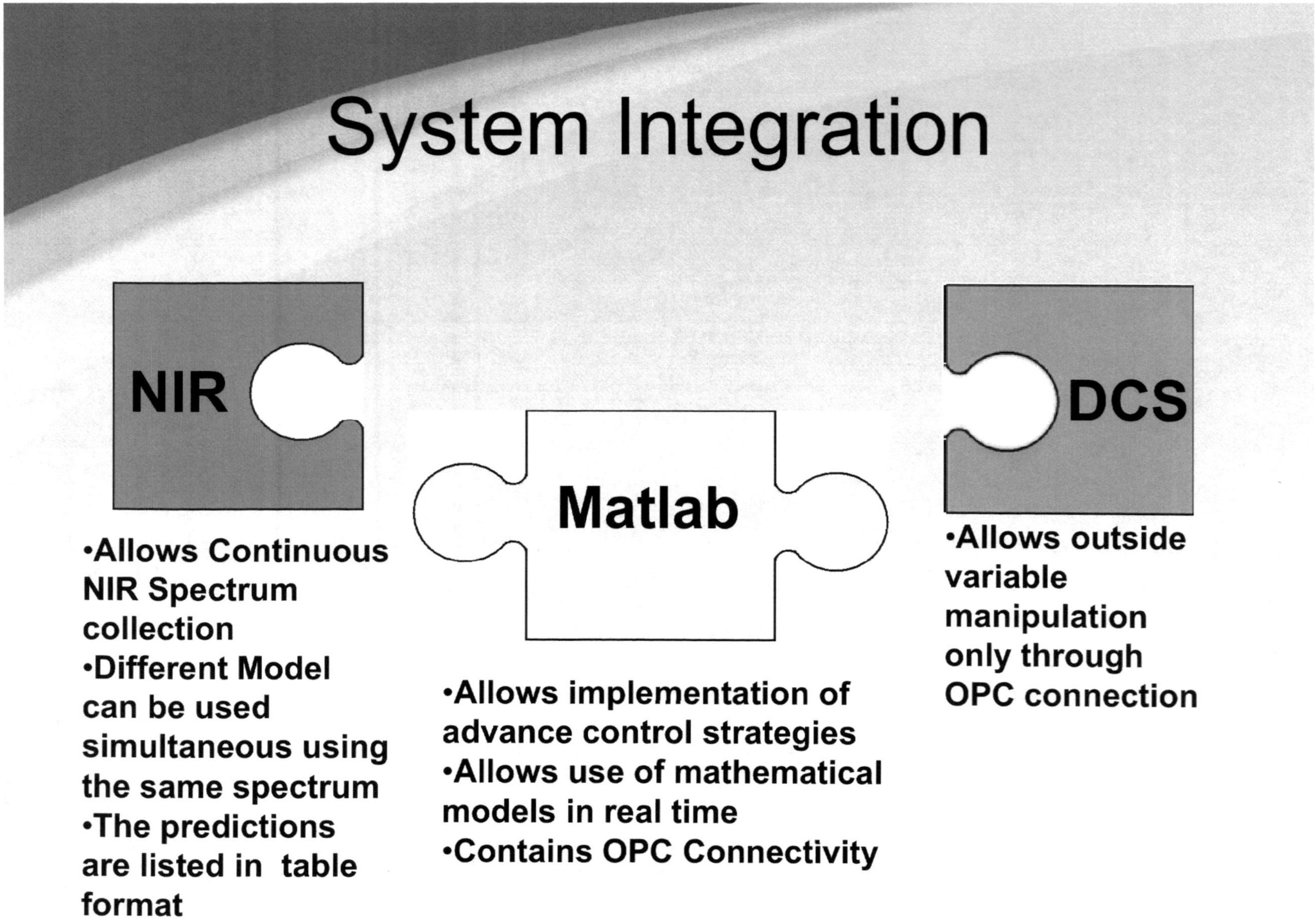
System Integration
NIR
Matlab
DCS
•Allows Continuous NIR Spectrum collection
•Different Model can be used simultaneous using the same spectrum
•The predictions are listed in table format
•Allows implementation of advance control strategies
•Allows use of mathematical models in real time
•Contains OPC Connectivity
•Allows outside variable manipulation only through OPC connection

# System integration

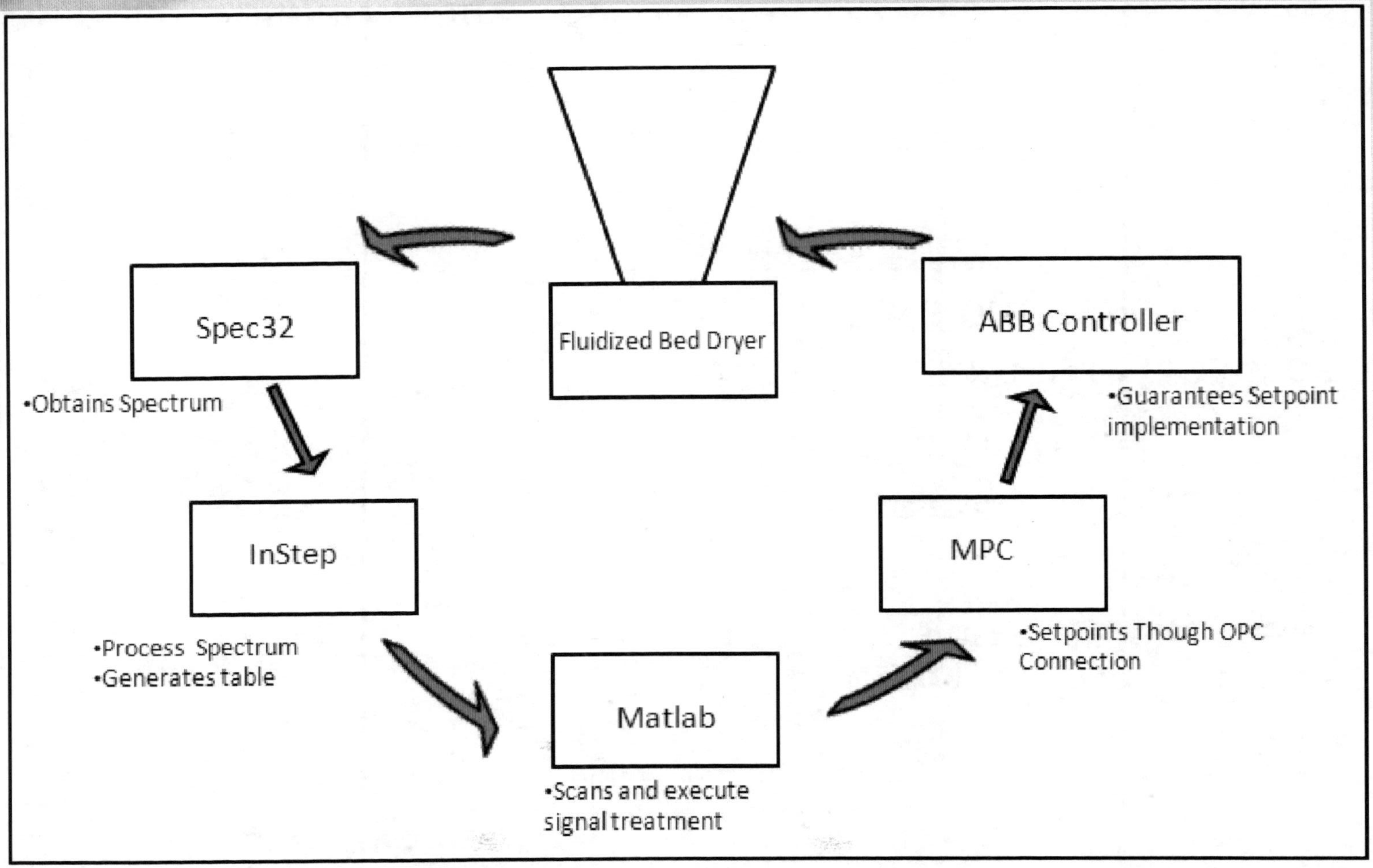

# Drying Process

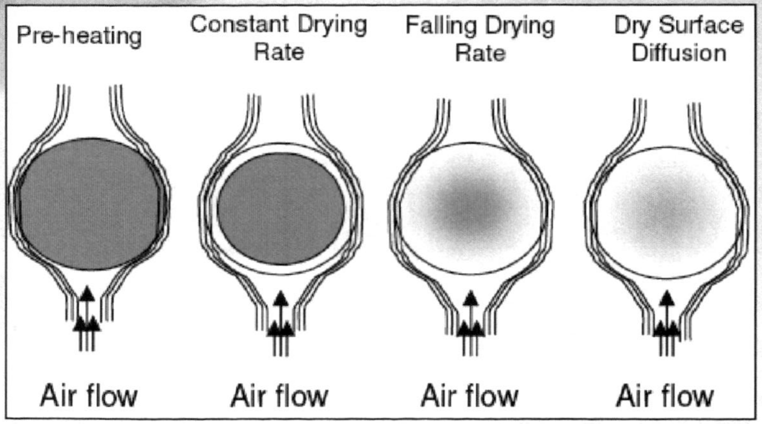


- The diffusion area changes as the particle dries up
- Depends on Particle Size Distribution (PSD)
- Drying becomes diffusive as process progresses

The final stages of the drying process is more diffusion driven allowing an optimization window for the process

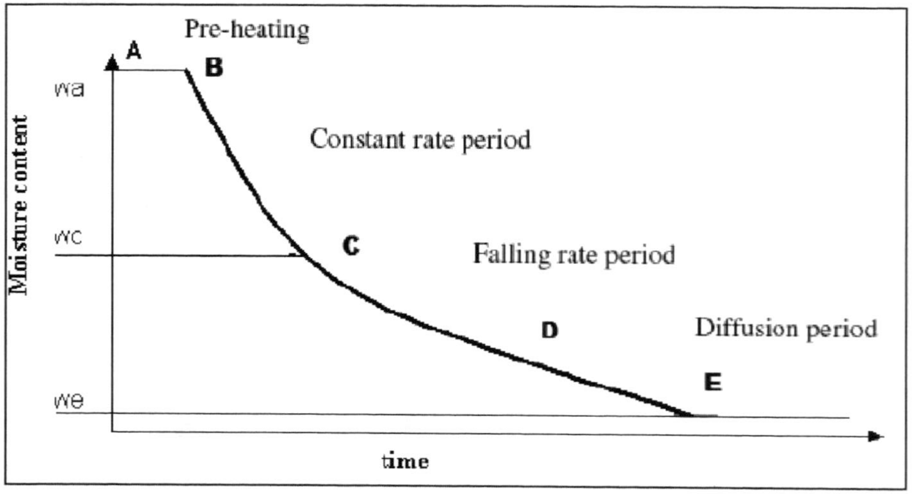

# Low Order Transfer Function: First Principal Model

- The First Principal Model was derived by Rivera*.
- Relates change in Powder Moisture content to the change in air Flux and air Temperature

$$\Delta Wp(t)=\left(Kq\Delta Q_{air,i}(t)+Kt\Delta T_i(t)\right)\left(1-e^{\frac{-t}{\tau}}\right)$$

*Wp = Moisture Content*

*Kq = Air Flux Proportionality constant*

*Kt = Temperature Proportionality constant*

*Q = Air Flux*

*T = Temperature*

*τ = Time Constant*

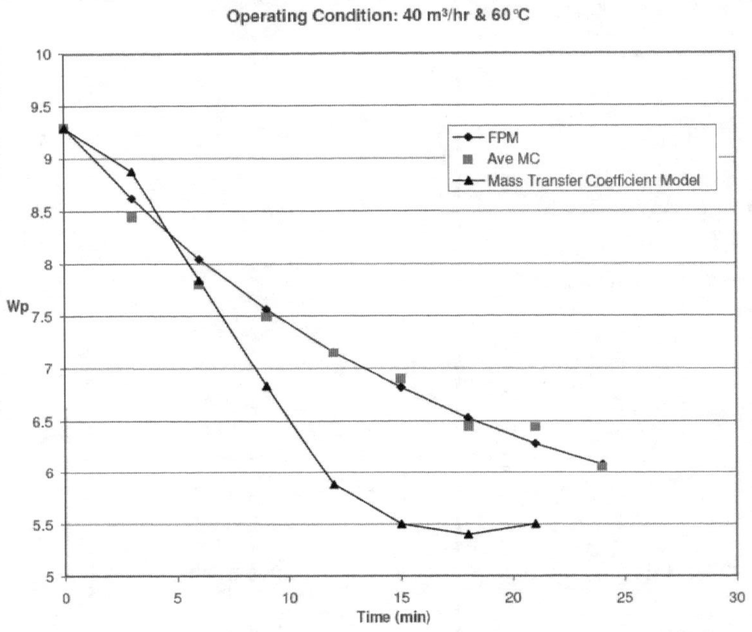


Rivera, J. O. Optimization of a Fluid Bed Dryer by the Implementation of a Model Predictive Controller. Thesis M.S. University of Puerto Rico, Mayaquez, P.R., (2006)

# FPM Parameter Determination

- The parameters are obtain using a fit tool on matlab.
- Assumptions
  - Step change at t=0
  - Constant Air Flux
  - Constant Temperature Differential

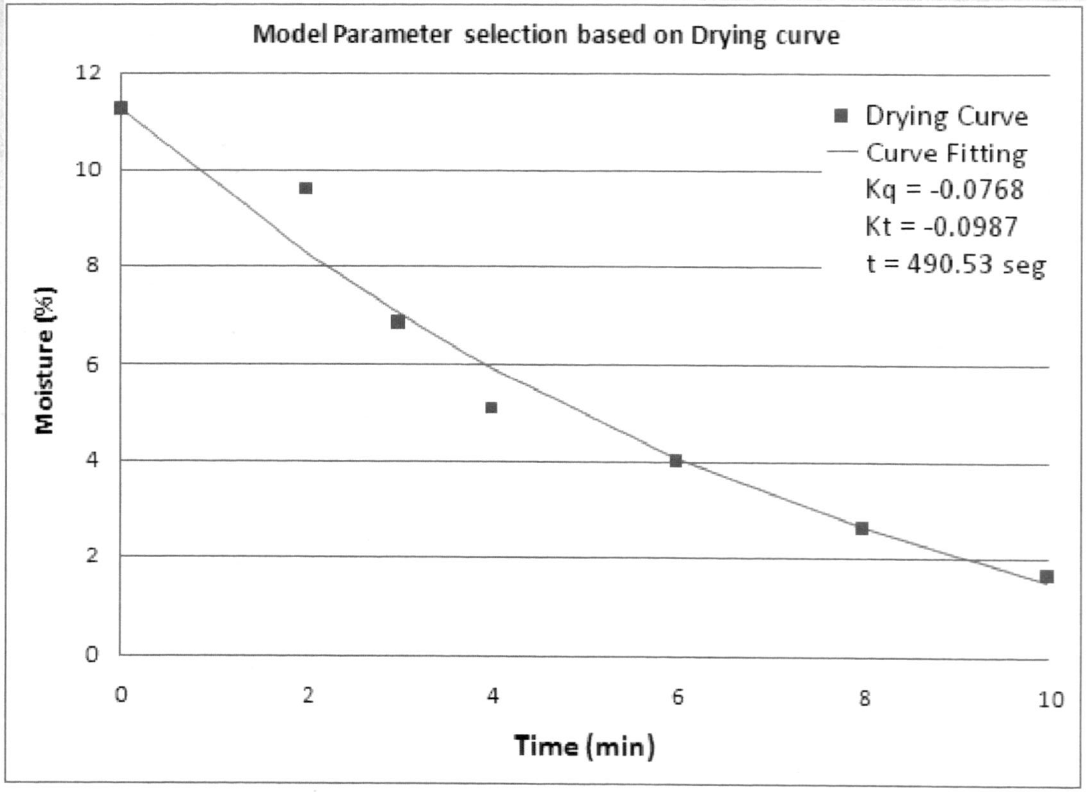

# Advanced Control Strategy: Model Predictive Controller*

- To optimize forecasts of process behavior
- Forecasting is accomplished with a process model
  - Allows process limitations
  - Determines optimum process parameters

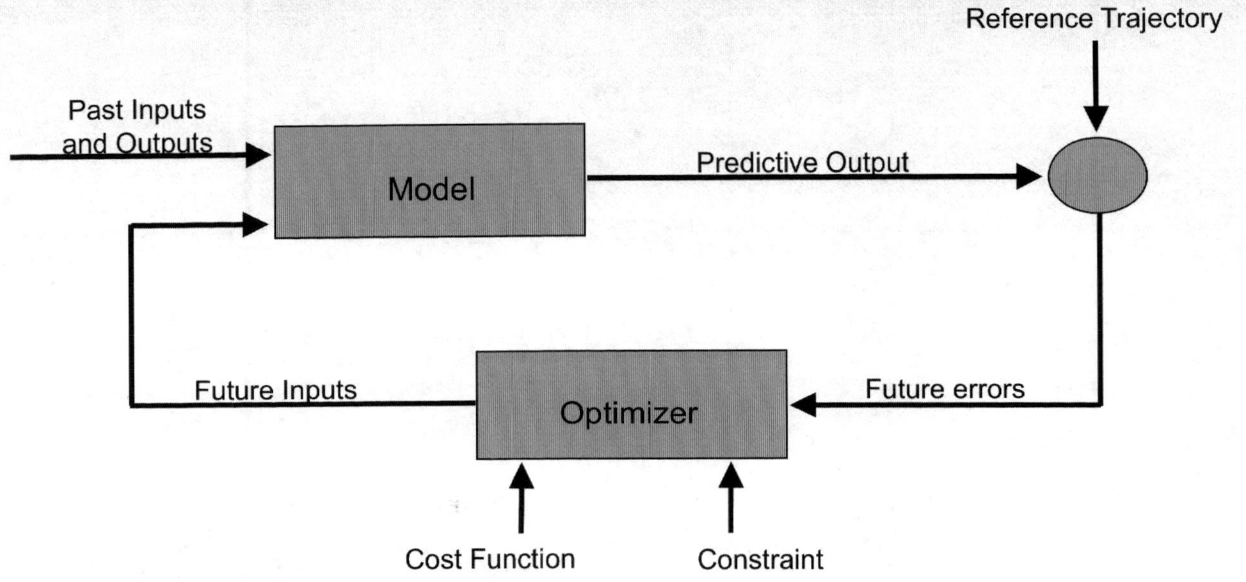


Rawlings J.B., *Tutorial Overview of Model Predictive Control*. IEEE Control Systems Magazine, pp.38-52, June 2000.

# MPC equations

System Model is changed into matrix form

$$\hat{\mathrm{W}}_{\mathrm{p}}(k+1) = K\Delta U(k) + \hat{\mathrm{W}}_{\mathrm{p}}(k)$$

Each component represents:

$$\mathrm{K} = \begin{bmatrix} K_1 & 0 & \cdots & 0 \\ 0 & K_2 & \ddots & \vdots \\ 0 & \ddots & \ddots & 0 \\ \vdots & \ddots & 0 & K_m \\ \vdots & \ddots & \vdots & \vdots \\ 0 & \cdots & 0 & K_p \end{bmatrix}$$

$$\mathrm{K}_{\mathrm{i}} = \left(1 - e^{-t_1/\tau}\right)\begin{bmatrix} K_q & K_T \end{bmatrix}$$

$$\Delta\mathrm{U(k)} = \left[u(k), u(k+1), \ldots, u(k+M-1)\right]^T$$

$$\mathrm{u} = \begin{bmatrix} Q_{air,i} & T_i \end{bmatrix}$$

# MPC equation

The performance index of the MPC controller

$$J = E(k+1)^T QE(k+1) + \Delta U(k)^T R\Delta U(k)$$

Solving the MPC controller for the control

$$\Delta \mathrm{U}(k) = (K^T QK + R)^{-1} K^T Q\hat{E}^{\circ}(k+1)$$

Which can be simplified as:

$$\Delta \mathrm{U}(k) = K_c \hat{E}^{\circ}(k+1)$$

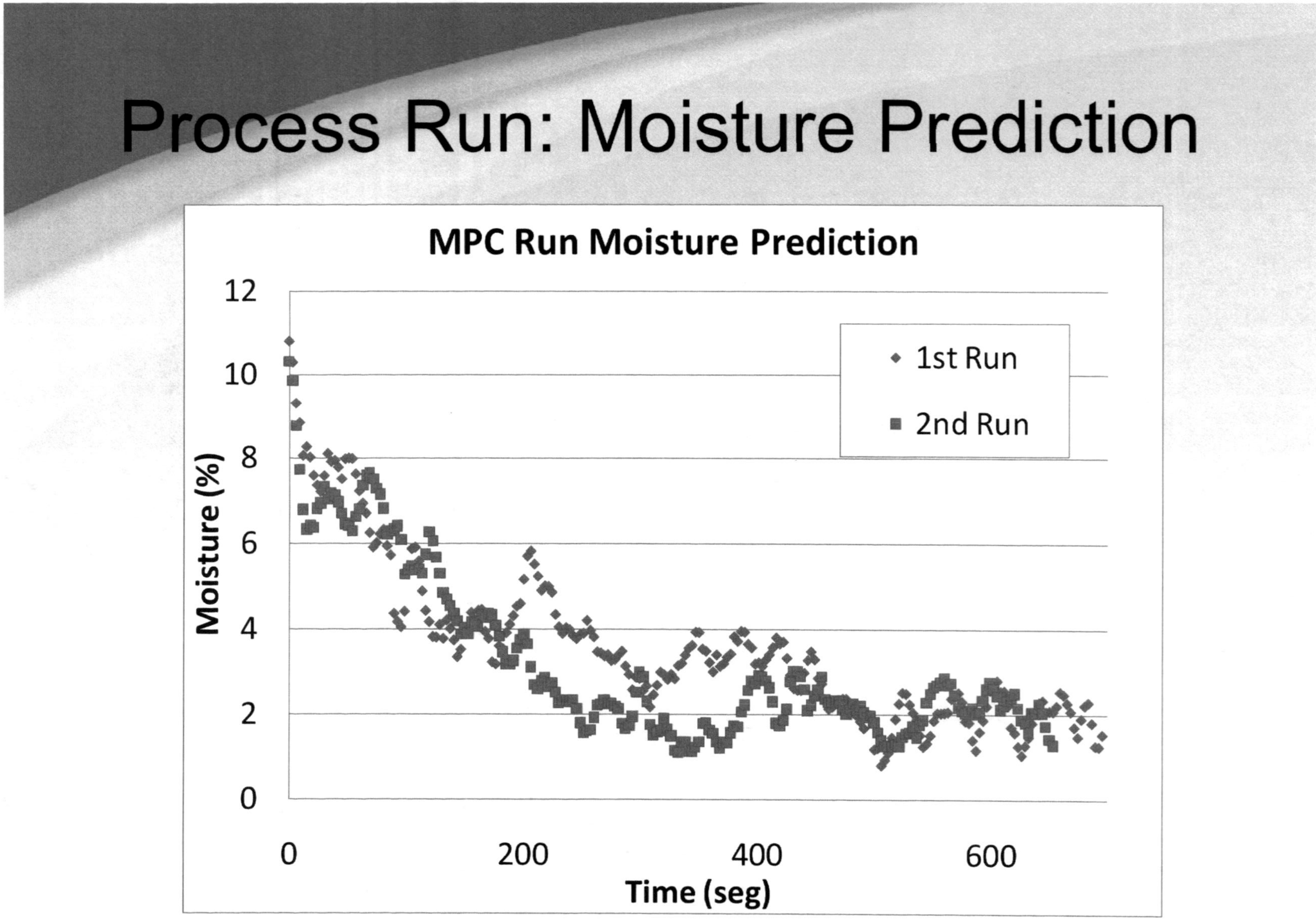
Process Run: Moisture Prediction
MPC Run Moisture Prediction
Moisture (%)
12
10
8
6
4
2
0
0
200
400
600
Time (seg)
1st Run
2nd Run

# Process run: set point and process variables

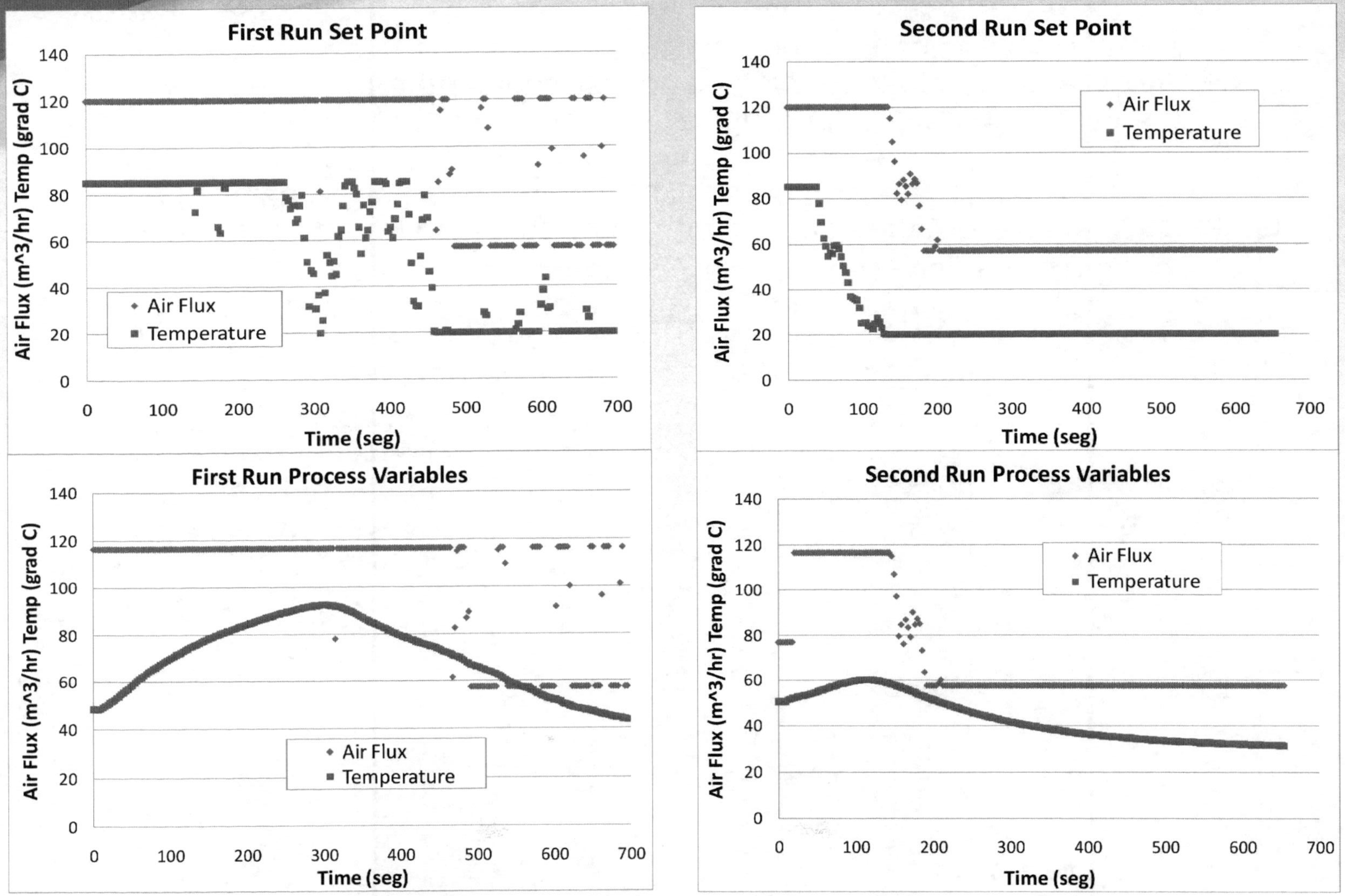

# References

Rivera, J. O. Optimization of a Fluid Bed Dryer by the Implementation of a Model Predictive Controller. Thesis M.S. University of Puerto Rico, Mayaguez, P.R., (2006)

J. Kumar, G. Warnecke, M. Peglow, S. Heinrich, Comparison of numerical methods for solving population balance equations incorporating aggregation and breakage, *Powder Technology* (2008), doi: 10.1016/j.powtec.2008.04.014

Rawlings J.B., *Tutorial Overview of Model Predictive Control*. IEEE Control Systems Magazine, pp.38-52, June 2000.

Boerefijn R., Hounslow M.J. Studies of fluid bed granulation in an industrial R&D context, Chemical Engineering Science (60) 2005 p 3879-3890

Rivera E. Optimization of Drying-End-Points Measurements for the Automation of a Fluidized-Bed Dryer Using FT-NIR Spectroscopy Thesis M.S. University of Puerto Rico, Mayaguez, P.R., (2004)

Roggo Y., Chalus P., Maurer L., Lema-Martinez C., Edmond A., Jent N., A review of near infrared spectroscopy and chemometrics in pharmaceutical technologies, Journal of Pharmaceutical and Biomedical Analysis Vol. XX (2007) pp.

# Acknowledgments

- God
- My Family
- Dr. Carlos Velazquez and research group
- ERC – SOPS
- Pfizer Pharmaceutical

# Integration of Principal Component Modeling Methods in a Model Predictive Control Architecture

**Chris McCready**, Process Analytical Technology, Umetrics Inc., Kinnelon, NJ

Manufacturing processes and measurement systems are becoming increasingly more complex. Recently it has become a common practice to use scores and residual statistics from principal component analysis (PCA) and partial least squares (PLS) models to summarize and monitor the state of a process. This session describes the integration of these PCS and PLS statistical metrics in a model predictive control (MPC) structure to provide the ability to maintain a process within a desired operating region. This is particularly relevant for the regulated pharmaceutical and biotech manufacturing environments where quality by design and design space quality systems require processes to be maintained within operating windows that assure quality.

See more of this Session: Automatic Control and Qbd
See more of this Group/Topical: Topical I: Comprehensive Quality by Design in Pharmaceutical Development and Manufacture

# A QbD Approach to Selection of Materials for Drug Product Packaging and Medical Devices

Arthur J. Shaw III, Analytical Research and Development, Pfizer Inc., Groton, CT

This presentation will describe a QbD approach to the selection of materials used to package drug products or to construct medical devices that deliver drug products. Special attention is required for materials that contact drug products due to extractables and leachables (E&L) considerations. Leachables, chemicals that migrate from the materials into the drug product, can cause safety toxicity concerns and/or stability issues from unwanted reactions with the drug product. Testing is required to demonstrate the quality of the drug product at end-of-shelf-life. To meet global supply requirements, a shelf-life period of 2 or more years is normally desired. A miss-step in the selection of packaging or device materials can lead to costly material change issues and delayed launch schedules if detection of unwanted leachables occur during late stage testing.

This presentation outlines a QbD E&L process for materials that are normally purchased for packaging or device purposes. The process can reduce the risk of poor choices in materials due to E&L issues. The presentation will include examples and will discuss industry initiates currently under development to help in the material selection, testing, and evaluation of extractables and leachables.

See more of this Session: QbD in Solid-Liquid Processes
See more of this Group/Topical: Topical I: Comprehensive Quality by Design in Pharmaceutical Development and Manufacture

# Application of Modeling to Scale-up Liquid Mixing Processes

**Venkat R. Koganti**, Parenteral Center of Emphasis, Pfizer Global Research and Development, Groton, CT

Fred Carroll, Pgrd, Pfizer Global Research and Development, Groton, CT

Richard Ferraina, Pgrd, Pfizer Global Research and Development, Groton, CT

Mark Berry, Pgrd, Pfizer Global Research and Development, Groton, CT

Robert Leasure, Pgm, Pfizer Global Research and Development, Groton, CT

Jeffrey Gaudio, Pgm, Pfizer Global Research and Development, Groton, CT

Yang Liu, Pgrd, Pfizer Global Research and Development, Groton, CT

Kenneth Norris, Pgrd, Pfizer Global Research and Development, Groton, CT

Bryan Braxton, Pgrd, Pfizer Global Research and Development, Groton, CT

WIlliam Petre, Pgrd, Pfizer Global Research and Development, Groton, CT

Yogesh Waghmare, Bend Research Inc, Bend, OR

Rick Falk, Bend Research Inc, Bend, OR

Scale-up of liquid mixing is usually done empirically without considering the several factors like dissolution theory, equipment differences between scales and so on. In this work we present a systematic way to scale-up liquid mixing processes utilizing both Design of Experiment (DoE) approach to identify the operating space (effect of operating parameters) and Computational Fluid Dynamics modeling to scale the mixing dynamics from laboratory to commercial scale. Underlying principles of modeling and also experimental data on dissolution at different scales for model solutes will be presented. This work also serves as an example of utilizing Quality by Design principles in designing the operating space during drug product manufacturing process design.

See more of this Session: QbD in Solid-Liquid Processes
See more of this Group/Topical: Topical I: Comprehensive Quality by Design in Pharmaceutical Development and Manufacture

# Application of Quality by Design Principles to Development of Freeze-Dried Parenteral Dosage Forms

Linas Mockus, Purdue University

Prabir K. Basu, Purdue University

Mansoor A. Khan, FDA

Nancy J. Harper, Pfizer, Inc.

Steven L. Nail, Baxter Pharmaceutical Solutions LLC

Nathan Pease, Baxter Pharmaceutical Solutions LLC

Tim Paul, Baxter Pharmaceutical Solutions LLC

Michael J. Pikal, University of Connecticut

David LeBlond, Abbott Laboratories

Pharmaceutical manufacturing is going through a major transition from quality assessment by testing the product at the end of manufacture to the more holistic control strategy of quality by design and Process Analytical Technology. There is greater realization that an enhanced understanding of the pharmaceutical processes involved in product manufacture and in particular, appreciation of criticality of the process steps and the critical process parameters that control each of these steps are of vital importance to ensuring quality of manufactured products.

The FDA and the industry strongly believe that through implementation of QbD, pharmaceutical manufacturing can reach its "Desired State" as adopted by the International Conference on Harmonization:

1. Product quality and performance achieved and assured by design of effective and efficient manufacturing processes
2. Product specifications based on mechanistic understanding of how formulation and process factors impact product performance
3. Ability to effect continuous improvement and continuous "real time" assurance of quality.

QbD is different from traditional quality activities which focus on improving existing products and processes using statistical process control. However, QbD refers more to design of new products or redesign of existing products as if they were being looked at anew.

Some steps have already been taken to understand how QbD can be implemented in an industry-wide manner for oral solid dosage forms. The goal of this project is to reach a better understanding of the concept of QbD in general, and how it can be applied to the most common parenteral manufacturing processes.

The approach followed in this study is shown in Figure 1. Key elements emphasized are:

- Focusing on critical product and process parameters
- Use of cause and effect matrices and Ishikawa diagrams for risk assessment
- Model-based simulations process optimization and design space construction
- Development of the "Design Space" based on prior knowledge, prediction, and experimental results
- Construction of design space that may not require 'edge of failure' (absence of trial and error)
- Linking control strategy to design space
- In-line process control

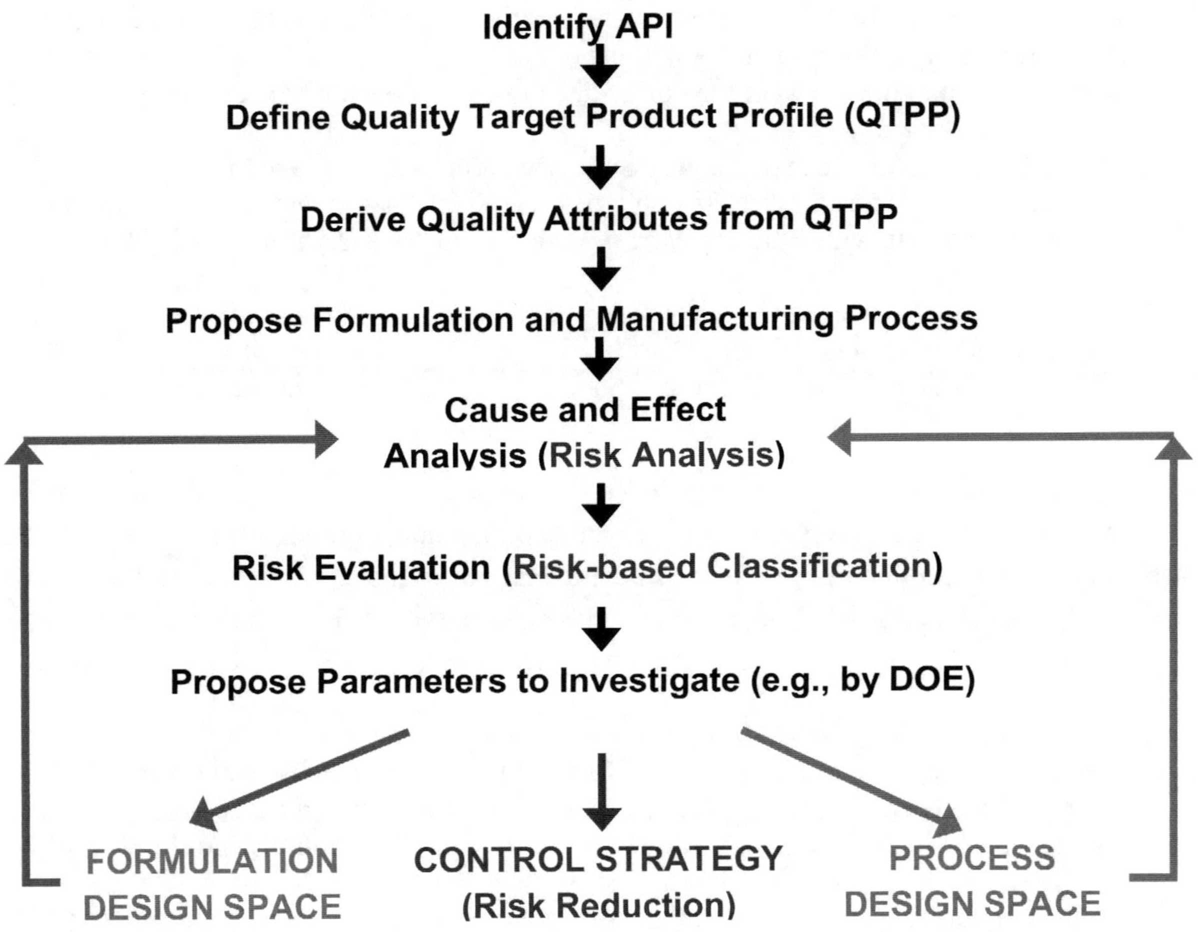


**Figure 1 QbD approach to developing a parenteral formulation for a "small molecule"**

Sodium ethacrynate, typical API which is widely available and representative of a small molecule parenteral drug, was selected for this study. Key degradation pathways, namely dimerization of drug and hydration of methylene double bond were verified as part of the project.

The quality by design (QbD) process followed is as stated below:

1. First, the desired performance attributes were selected via an exercise of identifying the quality target product profile. From this profile, an initial list of measurable critical quality attributes (CQA) was derived. In Six Sigma world this part of the QbD process is analogous to building a critical to quality (CTQ) tree where the quality target product profile represents customer requirements and critical quality attributes that are the key measurable characteristics of a product or process whose performance standards or specification limits must be met in order to satisfy the customer.
2. The pharmaceutical development was then conceptually subdivided into several focus areas from design to freeze-drying.
3. A risk assessment was performed to identify the variables and unit operations that most likely impact the CQAs for each focus area. Ishikawa diagrams were employed for the variable identification purposes.

4. The risk assessment also helped to establish functional relationships between key inputs (design/process parameters) and key outputs (CQAs).
5. Design/process parameters were then prioritized to focus development activities on potentially high-risk areas.
6. Cause and effect (C&E) matrix, another Six Sigma tool, was employed for risk assessment. There is a natural relationship between C&E matrix and failure mode and effects analysis (FMEA). FMEA may be used during the risk assessment activities as well.
7. Once the C&E matrix was defined, an experimental approach was assigned to all critical design/process parameters.
8. Based on experimental results and prior knowledge, the design space was defined.

The experimental step of the QbD process included formulation and process optimization. In general, the overall risk assessment process drove the experimental work. However, in some instances concurrent execution of risk assessment and experimental steps proved to be beneficial from a scientific insight and overall development time perspective. It should be noted that experimental results may sometimes drive revision of risk assessment. Some important steps in the process followed were:

1. API solubility was identified as a CQA. The solubility study, namely determining the effect of pH and temperature on solubility of pure drug, was started immediately. It was found that solubility, both at room temperature and at 5°C is essentially independent of pH in the range from 6 to 8. In addition, it was determined that solubility in water at room temperature is approximately 50 mg/mL, while at 5°C the solubility decreased to 20 mg/mL. With the aid of thermo-gravimetric analysis (TGA) it was determined that the compound is capable of forming dihydrate during precipitation from supernatant solution. However, only insignificant water loss was observed for the freeze-dried solid, indicating that the freeze-dried solid is predominantly anhydrate. The formation of dihydrate was also confirmed by X-ray powder diffraction. In both studies, prior knowledge about the drug was not adequate to make a sound scientific judgment and experimentation was deemed necessary.
2. Another CQA that required immediate attention was aqueous solution stability. Preliminary experimental studies in order to determine the effects of pH and buffer concentration on the aqueous solution stability were started well before the C&E matrix was completed and experimental strategy assigned. The study was performed over a two week period at room temperature. It was determined that lower buffer concentrations and lower pH caused less degradation in the aqueous solution. Consistent with the findings reported in the literature, it was established that the primary degradation pathway in aqueous solution involves hydration of methylene double bond in the presence of water. Preliminary studies on the effect of pH, buffer concentration, and annealing conditions on the stability of freeze-dried solid were also initiated even before the experimental strategy for the design focus area was formally defined. The study demonstrated that stability of drug increases with higher pH and lower buffer concentrations. The unusually stability behavior in aqueous and solid states is attributed to different degradation pathways. Consistent with the findings reported in the literature, we established that the primary

degradation pathway in the freeze-dried solid involves dimerization while in aqueous solution the hydration of methylene double bond is predominant. Unexpectedly, the annealing effect was insignificant. The take-away from those preliminary experiments is the need to fill the gaps in the scientific knowledge and actually facilitate risk assessment by providing more data.

3. One the most important CQAs identified was drug stability. To gain more experience about the effect of various common excipients and bulking agents on the stability of final dosage form, an initial characterization and formulation pre-screening was initiated. The formulations of pure drug in different concentrations, of drug with mannitol in different concentrations, and of drug with trehalose were considered. Based on the results of accelerated stability at 60ºC over 2 month period it was determined that pure drug is most stable.
4. The formulation optimization culminated with a study of the effects of API concentration, pH, buffer concentration, and buffer type on the stability of freeze-dried solid and crystallinity of the drug. Sodium phosphate and citrate were considered as possible buffers. Based on the results of accelerated stability study at 40ºC over one month duration, the effects of pH and API concentration as well as their interaction were determined to be significant. The effects of buffer concentration and buffer type were found insignificant. X-ray powder diffraction indicated, within the precision of the method, that stability and crystallinity were not correlated. It was established that only API concentration and pH are critical while the effects of buffer concentration and buffer type are insignificant. The optimum formulation was found to be 20 mg/mL API in 5mM sodium phosphate buffer at neutral pH.
5. The effects of critical process parameters, such as loading time, ramp rate to freeze temperature, and annealing conditions, were determined experimentally by employing response surface design. To determine the optimum primary drying conditions, a pseudo-steady state heat and mass transfer model was used. The model defines functional relationships between primary drying temperature, chamber pressure, and target product temperature. A residual moisture study was conducted to determine the optimum secondary drying duration.
6. The QbD process concluded with the identification of appropriate controls during manufacturing. Since drug degrades rather quickly in a liquid solution, the control identified was to limit the duration of sodium ethacrynate being in liquid state to 24 hours. The knowledge gained may have to be reassessed once more experience about the drug and associated risks is acquired.

By following the QbD process and focusing on critical to quality design/process parameters, we were able to design a formulation and develop a process to satisfy attributes of the quality target product profile. The proposed process systematically related mechanistic understanding of design/process parameters to drug product critical quality attributes. Multivariate experiments were conducted to understand product and process. It replaced the mainly empirical, one-variable-at-a-time pharmaceutical development method. Trial and error was minimized. A design space was established within which the manufacturing process may be safely adjusted as opposed to a traditional fixed manufacturing process.

# On the Optimization of the Cooling Profile in Crystallization Using the Design of Dynamic Experiments Approach

Andrew Hagger, Systems Research Institute and Chemical and Biological Engineering, Tufts University, Medford, MA

Andrew Fiodalis, Systems Research Institute and Chemical and Biological Engineering, Tufts University, Medford, MA

Christos Georgakis, Systems Research Institute and Chemical and Biological Engineering, Tufts University, Medford, MA

To control the size and shape of crystals during batch crystallization, one must either have an accurate crystal growth and nucleation model or run a series of experiments to find the optimal operating conditions. Since the former is a very intricate and time-consuming task, especially for new compounds, the later will be applied using a novel technique, Design of Dynamic Experiments (DoDE)(Georgakis, 2008, 2009), to determine the optimal operating conditions for a batch crystallizer that yield the desired size and shape distributions of the product crystals. The novelty of DoDE is that it is a model-free approach. Instead of using a fundamental model, the optimal operating conditions are found by performing a number of experiments in which the time dependent decision variable(s) is (are) varied according to a systematic set of dynamic signatures, often constrained between minimum and maximum allowed values. The results of the experiments are then used to calculate the optimal profile of the time-dependent decision variable(s). This method determines the optimum processing conditions while utilizing a minimum number of resources. In crystallization the most prevalent time dependent decision variables are the cooling rate and the antisolvent addition. In this presentation, we will discuss results showing the usefulness of DoDE to optimize the cooling profile of a two-liter batch crystallization unit equipped with a model predictive temperature controller to obtain the desired cooling profile(s) in each of the DoDE experiments as well as in the optimal one producing the desired distribution of crystal population characteristics. While the methodology presented here is applicable to any crystallization, the experimental data reported relate to the crystallization of potassium dihydrogen phosphate ($KH_2PO_4$), or KDP.

---

Georgakis, C. (2008). Dynamic Design of Experiments for the Modeling and Optimization of Batch Process.

Georgakis, C. (2009, July 12-15). A Model-Free Methodology for the Optimization of Batch Processes: Design of Dynamic Experiments. Paper presented at the IFAC Symposium on Advanced Control of Chemical Processes 2009, Istanbul, Turkey.

See more of this Session: QbD in Solid-Liquid Processes
See more of this Group/Topical: Topical I: Comprehensive Quality by Design in Pharmaceutical Development and Manufacture

*AIChE 2009 Annual Meeting QbD Topical Conference Proceedings Paper (**585e**):*

# Quality-by-Design (QbD): Examining the Impact of Process Variables on Particle Size Distribution for a Pharmaceutical Co-precipitation Process Using Real-Time Lasentec FBRM and PVM Process Monitoring and Design of Experiments (DOE) Approach

Huiquan Wu, Maury White, and Mansoor A. Khan
Division of Product Quality Research (DPQR, HFD-940), OTR/OPS/CDER/FDA
The FDA White Oak Campus, Life Science Building 64
10903 New Hampshire Ave, Silver Spring, MD 20993.

**Abstract**: Particle size distribution (PSD) is a critical quality attribute for many pharmaceutical unit operations, such as crystallization, granulation, and mixing. Various formulation and process factors could impact PSD to different extents of a particular unit operation. Therefore, it is important to know how those variables impact the PSD and how to optimize various variables to achieve a desired PSD. In this work, a pharmaceutical (naproxen) and a polymer (eudragit) co-precipitation process was monitored in real-time by using Lasentec FBRM and PVM technologies. The impact of various process variables such as co-precipitation slurry temperature, non solvent addition rate, and stirring rate on the PSD of the co-precipiate were examined systematically, using a design of experiments (DoE) approach. The results were interpreted and discussed in the context of particulate process engineering principles. The practical implication of the process knowledge gained for the establishment of operational process space was discussed under the framework of ICH Q8(R1).

**Keywords**: Quality-by-Design (QbD), Process Analytical Technology (PAT), Co-precipitation, Real-time Process Monitoring, Critical Process Variable, Interaction, Design Space.

*Corresponding author: Huiquan Wu, phone: 301-796-0022; email: Huiquan.wu@fda.hhs.gov

## 1. Introduction

Particle size distribution (PSD) is a critical quality attribute for many pharmaceutical unit operations, such as crystallization, granulation, mixing, and tabletting. Various formulation and process factors could impact PSD to different extents for a particular unit operation. Therefore, it is important to know how those variables impact the PSD and how to optimize various variables to achieve a desired PSD. These two aspects constitute an essential element of pharmaceutical process Quality-by-Design (QbD).

In pharmaceutical industry, crystallization is the main separation and purification process for the manufacturing of drug substances. Not only does crystallization affect the efficiency of downstream processes and processing ability (such as filtration, drying, and formulating, etc.), the efficacy and bioavailability of the drug can be dependent on the final crystal form and crystal size. Therefore, the ability to effectively control the crystal shape and crystal size distribution (CSD) is essential such that desired crystal shape and crystal size distribution (CSD) can be obtained. Accordingly, an accurate measurement of the CSD is extremely important for crystallization process control.

A conceptually similar yet technologically important pharmaceutical process is co-precipitation, which has been adapted to improve the dissolution characteristics of some water-insoluble drugs in the pharmaceutical drug delivery systems. Some of the formulation development works had been reported previously[1-2]. However, our understanding to co-precipitation process had been very limited. In recent years, research endeavor[3-4] has been made to explore using Process Analytical Technology (PAT)[5] tools and Quality-by-Design (QbD)[6-9] principles to gain process understanding of a pharmaceutical co-precipitation process.

In this work, naproxen and eudragit L 100 have been selected as the model drug and polymer for the co-precipitation study. Lasentec Focused Beam Reflectance Measurement (FBRM) and Particle Video Microscopy (PVM) techniques have been used to monitor the co-precipitation process in real-time, such that real time process data such as particle size distribution (PSD) and particle morphology could be obtained in situ. The purpose of this study is to study how process variables such as co-precipitation slurry temperature, non solvent addition rate, and stirring rate impact the particle size distribution (PSD) of the co-precipitate, using a design of experiments (DOE) and real time process monitoring approach.

## 2. Real-time Analysis of the Solid Phase: Lasentec FBRM

The measurement principle of the focused beam reflectance measurement is illustrated in Figure 1. A focused laser beam rotates at high speed and propagates into the particle suspension to be monitored. When the focused laser beam crosses a particle in front of the probe, a signal is backscattered into the probe. The length of the scanned chord is determined in the electronics of the system and transferred into a chord length distribution (CLD) histogram. Thus, the CLD provides online particle count and particle dimension information. Typically around 10000 particles are measured and counted each second providing a new CLD to track the dynamics of the ongoing particulate process. A CLD is a highly precise and sensitive means of tracking changes in both particle dimension and particle population. Also, no particle shape is assumed. The CLD is essentially unique for any given particle size and shape distribution. Among its broad applications, FBRM can be used to dynamically quantify and control the effect of process variables on a particulate system.

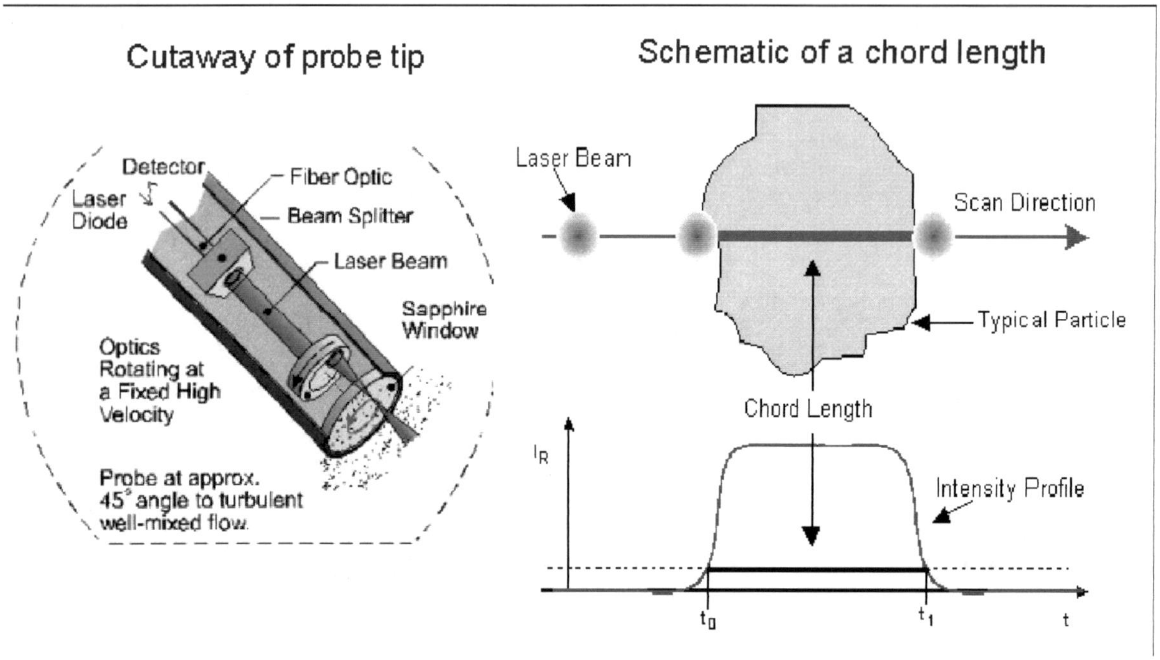


Figure 1: Focused beam reflectance measurement (FBRM): FBRM probe tip (left) and chord measurement (right): the laser beam direction is perpendicular to the paper. (figure provide by Mettler-Toledo Autochem, Inc.)

In this study, the FBRM D600 L system was set to collect CLDs with a laser speed of 2 m/s, applying the standard 'F-electronic' mode and using measurement duration of 2 seconds. A Lasentec PVM V819 system was used to collect the particle image in situ.

## 3. Experimental

3.1. Experimental details

Naproxen USP (lot No: NPX 368) was obtained from Albemarle Corporation (Orangeburg, SC). Eudragit L 100 (lot No: 1221203048) was obtained from

Röhm America Inc. (Somerset, NJ). Solvent (A) reagent alcohol (HPLC grade, lot No: 053546) was purchased from Fisher Scientific. Solvent B DI water was obtained from a FDA in-house facility and was kept in refrigerator at $3^0$C prior to use. All of these chemicals and solvents were used without any further processing or purification prior to the use for this experimental work.

The naproxen and eudragit were dissolved separately in 150 and 100 mL of Reagent Alcohol (Fisher Scientific Product Number A995-4 Lot 071844) respectively. These solutions were then added to the ChemGlass reaction vessel via a funnel. The stir speed was set on the OptiChem digital overhead stirrer. The temperature was set on the ThermoScientific Neslab RTE 740 thermoregulator. The system was allowed to reach the pre-defined temperature prior to the data acquisition. Once this temperature had been achieved, real-time monitoring began using the Lasentec FBRM system and PVM system, prior to the addition of any non solvent. After establishing a baseline, 125 mL of non solvent at about $3^0$C was added to the system at the pre-defined flow rate via a Cole-Parmer peristaltic pump. The system was allowed to come to equilibrium over a period of ten minutes, after which, an additional 50 mL of non solvent was added to the system. Depending on the temperature of the individual experiment ($15^0$C, $25^0$C, or $35^0$C ), the system was dosed an additional 3, 4, or 5 times (respectively) after the initial 125 mL dose. The system was allowed to reach equilibrium over a period of ten minutes after each individual dose.

## 3.2. Design of Experiments (DOE)

In this study, three process variables (slurry temperature ($^0$C), slurry stirring rate (RPM), non solvent addition rate (ml/sec) were selected as factors. Each factor was tested with three levels. A 3X3X3 full factorial design was created using JMP 7.0 (Cary, NC) software, as shown in Table 1.

Table 1 The 3X3X3 Full factorial design for the co-precipitation study

| Run No | Pattern | X1($^0$C) | X2(RPM) | X3(ml/sec) |
|---|---|---|---|---|
| 1 | 133 | 15 | 400 | 6.66 |
| 2 | 213 | 25 | 50 | 6.66 |
| 3 | 312 | 35 | 50 | 3.66 |
| 4 | 132 | 15 | 400 | 3.66 |
| 5 | 331 | 35 | 400 | 0.83 |
| 6 | 123 | 15 | 200 | 6.66 |
| 7 | 323 | 35 | 200 | 6.66 |
| 8 | 221 | 25 | 200 | 0.83 |
| 9 | 313 | 35 | 50 | 6.66 |
| 10 | 222 | 25 | 200 | 3.66 |
| 11 | 321 | 35 | 200 | 0.83 |
| 12 | 212 | 25 | 50 | 3.66 |
| 13 | 311 | 35 | 50 | 0.83 |
| 14 | 131 | 15 | 400 | 0.83 |
| 15 | 211 | 25 | 50 | 0.83 |
| 16 | 332 | 35 | 400 | 3.66 |
| 17 | 223 | 25 | 200 | 6.66 |
| 18 | 112 | 15 | 50 | 3.66 |
| 19 | 122 | 15 | 200 | 3.66 |
| 20 | 231 | 25 | 400 | 0.83 |
| 21 | 232 | 25 | 400 | 3.66 |
| 22 | 121 | 15 | 200 | 0.83 |
| 23 | 233 | 25 | 400 | 6.66 |
| 24 | 322 | 35 | 200 | 3.66 |
| 25 | 113 | 15 | 50 | 6.66 |
| 26 | 111 | 15 | 50 | 0.83 |
| 27 | 333 | 35 | 400 | 6.66 |

**4. Results and Discussion**

4.1. Typical Profile of CLD vs. Time

Various process variables could impact both the thermodynamics and the kinetics of a co-precipitation process to various extents, depending on the combination of process conditions. Theoretically, co-precipitation slurry temperature could directly impact the super-saturation of the co-precipitation system. Slurry stirring rate could directly impact the concentration distributions of the drug and polymer, thus impact the mass-transfer inside the co-precipitation vessel. For non-solvent addition rate, on the one hand, it directly impacts the creation of super-saturation inside the co-precipitation vessel; on the other hand, it indirectly impacts the re-distribution of various species inside the co-precipitation vessel. Their impacts on super-saturation, mass transfer, and heat transfer will be ultimately reflected by changes of process thermodynamics and process kinetics, including nucleation and co-precipitation growth, as illustrated by the data analysis results below.

Co-precipitation process rate was derived from the slope of CLD vs. time. A typical profile of CLD vs. time was presented in Figure 2. The non solvent was dosed three times with a time interval of approximately 10 minutes. A relative small increase of counts was observed for the initial dose. A large increase of counts was observed for the $2^{nd}$ dose. However, the increases of counts at various particle size ranges were not uniform. Instead, an increase rate dispersion phenomenon was easily appreciated. The most dominant increase occurs for the particle size range of [19.95 μm, 50.12 μm]; the $2^{nd}$ largest increase occurs for the particle size ranges of both [10 μm, 19.5 μm] and [50.12 μm, 100 μm]. For the $3^{rd}$ dose, only a small increase of counts was observed. This increase rate dispersion phenomenon was observed almost universally for most of the rest 26 batches with different combinations of process variables.

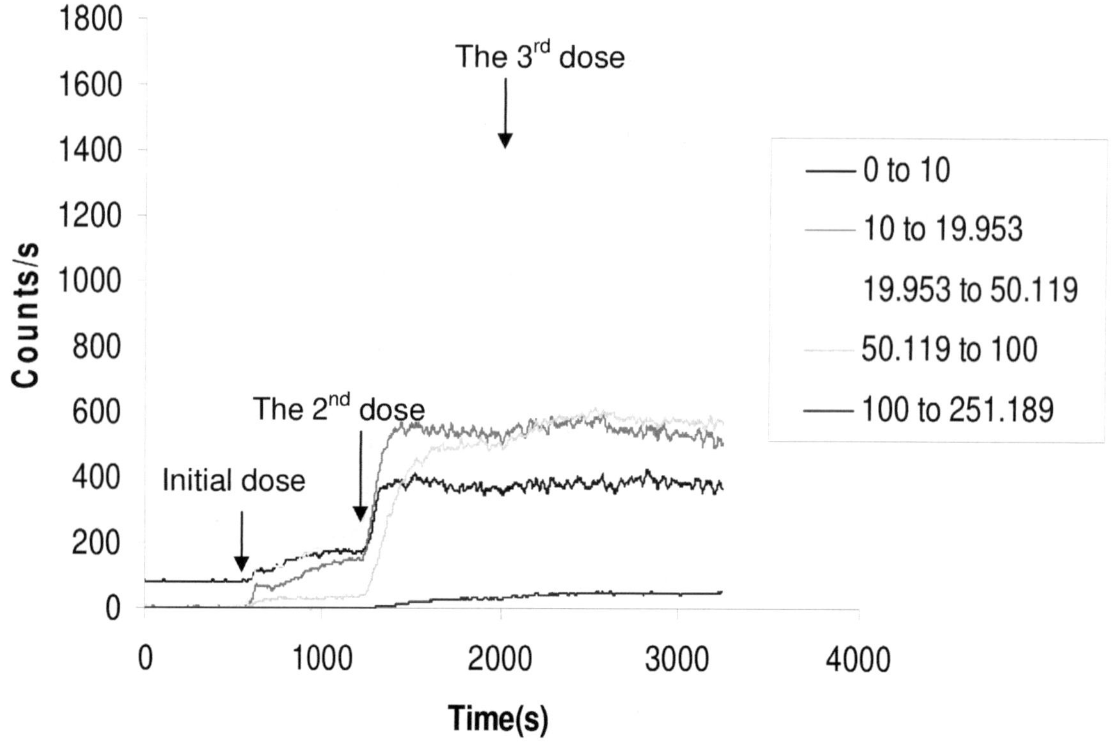


Figure 2 A typical profile of CLD vs. time for batch run F1-15-50-1.

4.2. Statistical Data Analysis

The slopes of CLD counts vs. time at each particle size ranges were calculated based on the profile of CLD vs. time. Statistical data analysis was performed to determine if there are any relationships between the process variables and the slopes (CLD increase rates) at particle size ranges between [[0 μm, 100 μm]. Results were shown below.

*4.2.1. Screening for most significant variables and interactions*

Table 2 shows that among the three process variables, 6 two-way interactions, and 7 three-way interactions studied, co-precipitation slurry temperature (X1), non solvent addition rate (X3), interaction between co-precipitation slurry temperature (X1) and slurry stirring rate (X2), are statistically significant process variables and interaction term at significant level of 0.1. The Half Norm Plot in Figure 4 provides additional evidence to support this conclusion.

Table 2 Screening for slope at PS range of [[0 μm, 10 μm] for the initial dose

| **Term** | **Contrast** | **Plot of t-Ratio** | **Individual p-Value** |
|---|---|---|---|
| X1 | -2.70101 | | **0.0516** |
| X3 | 2.48877 | | **0.0690** |
| X2 | -1.03939 | | 0.4148 |
| X1*X1 | 0.94202 | | 0.4644 |
| X1*X3 | -1.53676 | | 0.2354 |
| X3*X3 | -0.80989 | | 0.5627 |
| X1*X2 | 2.28029 | | 0.0918 |
| X3*X2 | 0.05323 | | 0.9688 |
| X2*X2 | 0.56050 | | 0.6902 |
| X1*X1*X3 | 1.11059 | | 0.3826 |
| X1*X3*X3 | 0.24734 | | 0.8602 |
| X1*X1*X2 | 0.78626 | | 0.5738 |
| X1*X3*X2 | 1.28132 | | 0.3128 |
| X3*X3*X2 | 2.04330 | | 0.1245 |
| X1*X2*X2 | 0.07383 | | 0.9572 |
| X3*X2*X2 | 0.01705 | | 0.9899 |

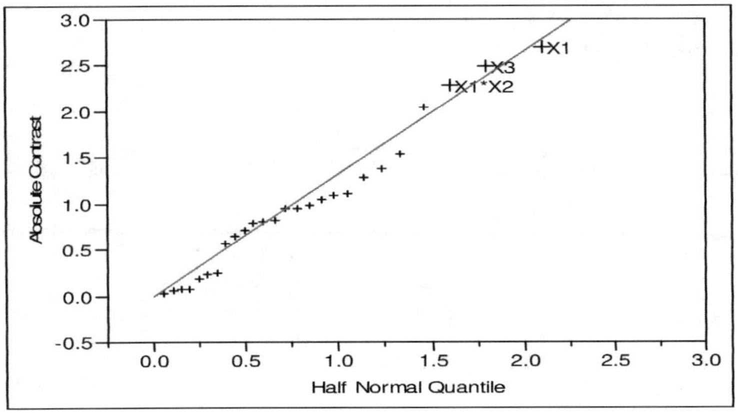


Figure 4. Half Norm Plot for screening results in Table 2

*4.2.2. Quantitative correlations between process variables and slopes (CLD increase rates)*

Taking data in Table 2 as one example, an attempt was made to build a quantitative model to describe the correlation between the significant variables and interaction with the slope data or the CLD increase rates at PS range of [[0 μm, 10 μm] for the initial dose. The results were shown in Figure 5. The fit line has a $R^2$ value of 0.52. Although this $R^2$ value is not perfect, the quantitative correlation is evident at the least. Similar correlations were found between process variables and slope data or CLD increase rates at other PS ranges such as [10 μm, 19.5 μm], [19.95 μm, 50.12 μm], and [50.12 μm, 100 μm].

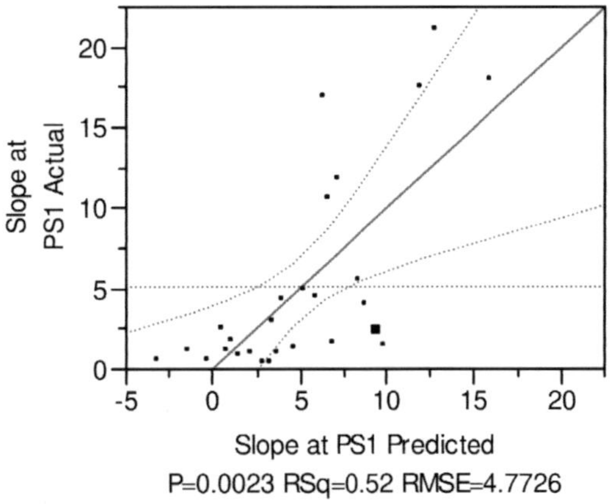


Figure 5. Least squares fit results: response slope at PS1 vs. actual slope at PS1.

4.3. Discussion from a process engineering perspective and regulatory science perspective

It is well known that crystallization process is a multi-phase process where mass transfer, heat transfer, and moment transfer phenomena could be coupled together to make the process control task very challenging. Pharmaceutical co-precipitation process, as being demonstrated in earlier work, is a complicated multi-phase process as well. As demonstrated in this work, process knowledge and in-depth process understanding are essential to establish appropriate process control strategy and operational space, such that co-precipitation product with desired quality attributes and properties (such as PSD, purity, morphology, dissolution characteristics, etc.) could be achieved eventually. Therefore, they are critical elements for developing appropriate process design space, as defined in the relevant regulatory guidelines such as ICH Q8, Q8(R1), etc. This work as a first step represents our initial effort in the direction to create case studies which reflect at least some of the key elements in FDA PAT Guidance and the ICH guidelines such as Q8, Q8(R1), Q9, and Q10, using PAT tools and QbD principles to obtain process understanding and develop an appropriate design space.

## 5. Conclusions

In this work, a pharmaceutical (naproxen) and a polymer (eudragit) co-precipitation process was monitored in real-time by using Lasentec FBRM and PVM technologies. The impact of various process variables such as co-precipitation slurry temperature, non solvent addition rate, and stirring rate on the CLD of the co-precipiate were examined systematically, using a design of experiments (DoE) approach. The results were interpreted and discussed in the context of particulate process engineering principles. It was found that the CLD increase rate dispersion being universal phenomenon in the scope of this study. Statistical significant process variables and interactions were identified through statistical data analysis of the DOE results. Quantitative correlations were found between process variables and the CLD increase rates at various particle size ranges from 0 μm to 100 μm.

**Disclaimer**

*The views and opinions expressed in this paper are only those of the authors, and do not necessarily reflect the views or policies of the FDA.*

**Acknowledgements:**

Mr. Eric Dycus and Mr. Ben Smith from the Mettler Toledo AutoChem Inc. are acknowledged for their technical supports to our Lasentec FBRM and PVM systems.

## 6. References

1. Khan MA, *et al*. 1994. Optimization of process variables for the preparation of ibuprofen coprecipitates with Eudragit S100.Int J Pharm 102:185-192.

2. Zaghloul AA, Faltinek J, Vaithiyalingam SR, Reddy IK, and Khan MA. 2001.Naproxen-Eudragit microspheres : screening of process and formulation variables for the preparation of extended release tablets, Pharmazie 56(4):321-324.

3. Wu H and Khan MA. 2006. Mapping the State of a Pharmaceutical Co-precipitation Process: An Integrated Process Analytical Technology (PAT) Approach. Conference Proceedings of the 2006 AIChE Annual Meeting (CD-ROM). AIChE Proceeding P-235, ISBN:0-8169-1012-X.

4. Wu H and Khan MA. 2009. Quality-by-Design (QbD): An Integrated Process Analytical Technology (PAT) Approach for Real-time Monitoring and Mapping the State of a Pharmaceutical Co-precipitation Process. Accepted for publication by the Journal of Pharmaceutical Science (August 2009).

5. FDA, 2004. "Guidance for Industry. PAT—A Framework for Innovative Pharmaceutical Development, Manufacturing, and Quality Assurance". Available at:

6. FDA/ICH, 2006. Guidance for Industry. Q8 Pharmaceutical Development. Available at: http://www.fda.gov/cder/guidance/6746fnl.pdf

7. FDA/ICH, 2008. Q8(R1) Pharmaceutical Development Revision 1. Draft available at: http://www.fda.gov/cder/guidance/8084dft.pdf

8. FDA/ICH, 2006. "Guidance for Industry. Q9 Quality Risk Management". Available at: http://www.fda.gov/cder/guidance/7153fnl.pdf

9. FDA/ICH, 2007. Q10. Pharmaceutical Quality System. Draft Consensus Guideline. Available at: http://www.fda.gov/cder/guidance/7891dft.pdf

# The *in Situ* Analysis and Characterization of the Coprecipitation of a Model Drug and Polymer

**Juandria V. Williams**, Cder/Ops/Otr/Dpqr, Food and Drug Administration, Silver Spring, MD

Rakhi B. Shah, Cder/Ops/Otr/Dpqr, Food and Drug Administration, Silver Spring, MD

Huiquan Wu, Cder/Ops/Otr/Dpqr, Food and Drug Administration, Silver Spring, MD

Mansoor A. Khan, Cder/Ops/Otr/Dpqr, Food and Drug Administration, Silver Spring, MD

Coprecipitation is a solid dispersion technique whereby a drug substance and polymer simultaneously precipitate out of solution. The technique has been adopted to enhance the bioavailability of poorly water-soluble drugs. While this process helps meet the need for improved drug delivery systems, it has not been well-studied to the point where process variability can be understood and controlled. "Quality by design" principles provide us the opportunity to understand and control the variability, such as formulation variables and particle growth dynamics, during the coprecipitation process. Thus, we can achieve desirable drug release profiles.

In this work, we designed and characterized a co-precipitation process, using a design of experiment strategy, observing which formulation variables, such as solvent type, anti-solvent addition rate and drug/polymer ratio, have the greatest affect on drug product. We then implemented two in situ technologies, FBRM and PVM to elucidate the effects of crystal nucleation and growth during coprecipitation, on the properties of the drug substance, excipient and drug/excipient complex. The metastable zone width was calculated and interpreted, such that the regions of optimum process conditions were identified.

See more of this Session: QbD in Solid-Liquid Processes
See more of this Group/Topical: Topical I: Comprehensive Quality by Design in Pharmaceutical Development and Manufacture

# Improving Process Understanding through Good Experimental Planning

Kelly Canter, Pharmaceutical Sciences, Pfizer Inc., St. Paul, MN

Leonard Perry, University of San Diego, San Diego, CA

The pharmaceutical industries ability to have manufacturing flexibility given they can demonstrate proper process understanding has brought a resurgence in the application of statistically designed experiments. Even though Design of Experiments (DoE) has been used extensively across all industries to characterize and optimize a manufacturing process, it is all too often their application falls short of its objective. Given the large emphasis to define and prove the manufacturing design space, it is imperative that the pharmaceutical industry use a planned and systematic approach to study the relationship between process parameters affecting those quality attributes critical to patient safety and efficacy. The selection of the proper experimental design requires correctly identifying the problem, associated factor ranges, and accompanying constraints. A large amount of literature exists on the theory and application of DoE, however minimal attention is given to the planning and sequential approach required to achieve the desired outcome. The focus of this paper is to provide guidelines for appropriate experimental planning, and a detailed summary of existing DoE and how their sequential application can support the development of the products design space and an efficient biopharmaceutical process.

See more of this Session: 11 Empirical Modeling and Experimental QbD
See more of this Group/Topical: Topical I: Comprehensive Quality by Design in Pharmaceutical Development and Manufacture

# The Use of the Design of Dynamic Experiments Methodology to Quantify the Design Space in QbD Tasks

Christos Georgakis, Systems Research Institute and Chemical & Biological Engineering, Tufts University, Medford, MA

The present communication describes a systematic methodology for the quantification of the Design Space characteristics of batch pharmaceutical process. The methodology is applicable to processes which are not known in great detail for a mathematical model to be available. The approach designs a set of experiments utilizing a newly developed methodology for the Design of Dynamic Experiments (DoDE) and the traditional response surface modeling methodology. The use of these input output-models enables the quantitative address the following issues related to the Quality by Design approach:

1) For processes in which no feedback control strategy is exercised:

1a. For a given variability in the raw materials and/or the processing conditions, what is the expected variability in the quality of the product produced?

1b. Given the maximal acceptable variability in the product and the current operating conditions, what is the maximum variability in the raw material that can be utilized without impacting the quality of the product?

2) For processes with a feedback control strategy on the expected product quality:

2a. What is the maximum variability in the raw materials or other uncontrolled disturbances which can be compensated by a given control strategy so the product quality is kept within acceptable limits?

2b. What is the impact of different fee-forward or/and feedback control strategies on question 2a?

Processes for which we have a model describing the cause and effect relationships among the pertinent variables can be also handled in a similar, if not easier, manner by substituting the design of dynamic experiments (DoDE) with an equivalent design of (selective) dynamic calculations (DoDC).

See more of this Session: 11 Empirical Modeling and Experimental QbD
See more of this Group/Topical: Topical I: Comprehensive Quality by Design in Pharmaceutical Development and Manufacture

# The ICH Q8 Definition of Design Space:
# A Comparison of the Overlapping Means and the Bayesian Predictive Approaches

by

John J. Peterson* and Kevin Lief

GlaxoSmithKline Pharmaceuticals

* E-mail: john.peterson@gsk.com

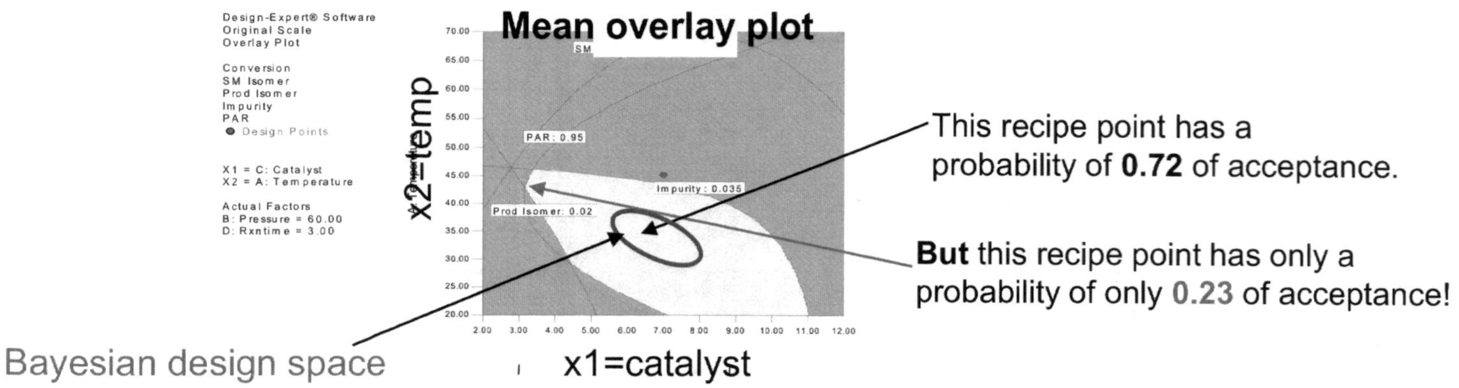

# ICH Q8 Definition of Design Space

The ICH Q8 FDA Guidance for Industry defines "Design Space" as: "The multidimensional combination and interaction of input variables (e.g. material attributes) and process parameters that have been demonstrated to provide assurance of quality.“

# ICH Q8 Definition of Design Space

The ICH Q8 FDA Guidance for Industry defines "Design Space" as: "The multidimensional combination and interaction of input variables (e.g. material attributes) and process parameters that have been demonstrated to provide assurance of quality.“

- *Three key concepts:*

1. **Measurement**
   For example: controllable factors, input material attributes, in-process measurements, quality response measurements.

# ICH Q8 Definition of Design Space

The ICH Q8 FDA Guidance for Industry defines "Design Space" as: "The multidimensional combination and interaction of input variables (e.g. material attributes) and process parameters that have been demonstrated to provide assurance of quality."

- *Three key concepts:*

1. **Measurement**
   For example: controllable factors, input material attributes, in-process measurements, quality response measurements.

2. **Prediction**
   - Models to relate the measurements to the relevant quality responses. These need to be compared to specifications for quality.
   - Mean predictions are not enough! Predictive distributions are necessary.

# ICH Q8 Definition of Design Space

The ICH Q8 FDA Guidance for Industry defines "Design Space" as:
"The multidimensional combination and interaction of input variables (e.g. material attributes) and process parameters that have been demonstrated to provide assurance of quality."

- *Three key concepts:*

1. **Measurement**
   For example: controllable factors, input material attributes, in-process measurements, quality response measurements.

2. **Prediction**
   - Models to relate the predictive measurements to the quality responses. These need to be compared to specifications for quality.
   - Mean predictions are not enough! Predictive distributions are necessary.

3. **Reliability**
   To quantify "How much assurance?"

- The QbD-oriented guidance (PAT, ICH Q8, Q9, Q10, etc) is inundated with the words "risk" and "risk-based".)
- See presentation by H. Gregg Claycamp (CDER), "Room for Probability in ICH Q9"

# Prediction and Reliability

1. Empirical or mechanistic models typically are used to make predictions about future response values for a specified set of process conditions.

2. Any complex process that is shut down and restarted from scratch (using different batches of starting materials, etc.) to produce a true replicate response value, will produce **different** responses even under "identical" operating conditions (even if using extremely accurate measuring devices).

3. So point 2 (above) implies that we do **not** have the following simplistic model concept:

"predicted response"= "true model function" + "measurement error".

# Prediction and Reliability

2. Any complex process that is shut down and restarted from scratch (using different batches of starting materials, etc.) to produce a true replicate response value, will produce **different** responses even under “identical” operating conditions (even if using extremely accurate measuring devices).

3. So point 2 (above) implies that we do **not** have the following simplistic model concept:

“predicted response”= “true (deterministic) model function” + “measurement error”.

4. Instead, we have the follow model concept:

“<u>distribution</u> of predicted responses” = “stochastic process” + “measurement error”.

# Prediction and Reliability

- **Predictive** *Distributions* are an important concept as they can be used to quantify the **Reliability** of meeting quality specifications.

- This is important "…to provide assurance of quality" as required by the ICH Q8 definition of design space.

  The ICH Q8 FDA Guidance for Industry defines "Design Space" as:

  "The multidimensional combination and interaction of input variables (e.g. material attributes) and process parameters that have been demonstrated to provide assurance of quality."

- ICH Q8 definition also begs the question: **"How much assurance?"**

## *How much assurance do we have of meeting product specifications?*

- For a given set of manufacturing conditions (i.e. a given "recipe"), there will be batch-to-batch variation.

- In theory, for many manufactured batches, there is a distribution of possible results.

- Ideally, we would like most all of this distribution to be well inside a region of product acceptance.

# How much assurance do we have of meeting specifications?

- Ideally, we would like most all of this distribution to be well inside a region of acceptance.

- Consider the specification regions below (in gray) for two different recipes:

Distribution of possible results for a recipe #1

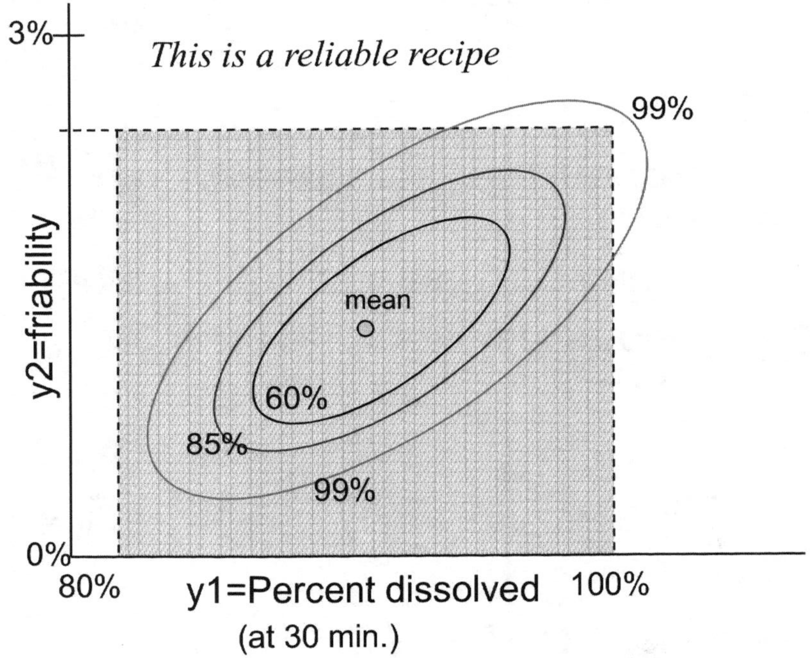


Distribution of possible results for recipe #2

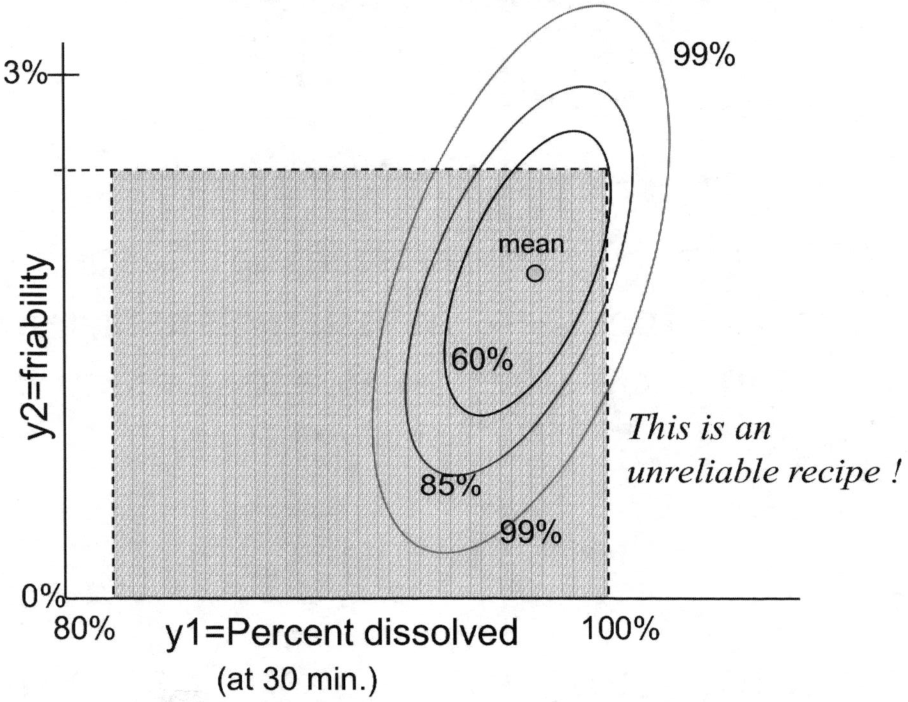

# How much assurance do we have of meeting specifications?

- The mathematics of “assurance” or “likelihood” is based upon the laws of probability.

- The laws of probability show that the proportion of a distribution inside of a region of acceptance provides the probability of obtaining an acceptable result.

Distribution of possible results for a recipe #1

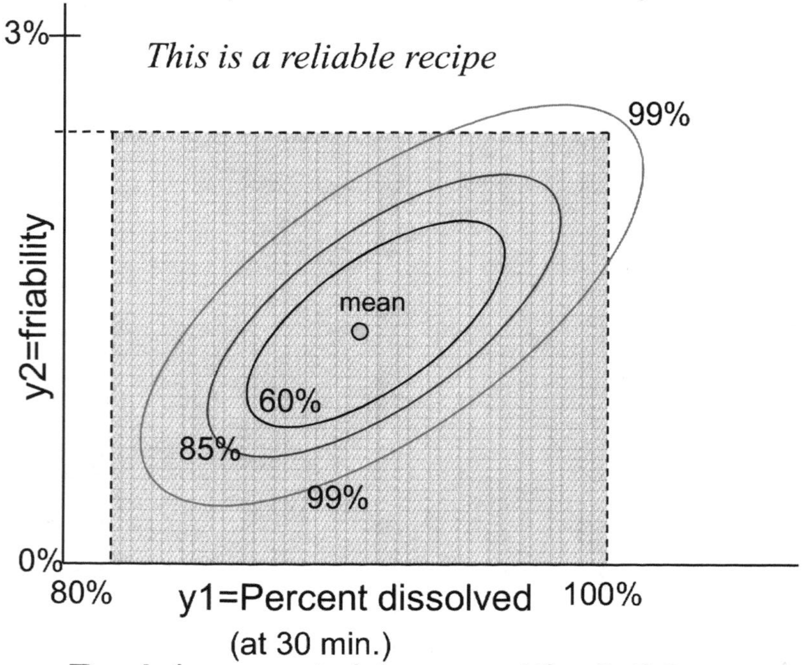


Prob(acceptable result)=0.90

Distribution of possible results for recipe #2

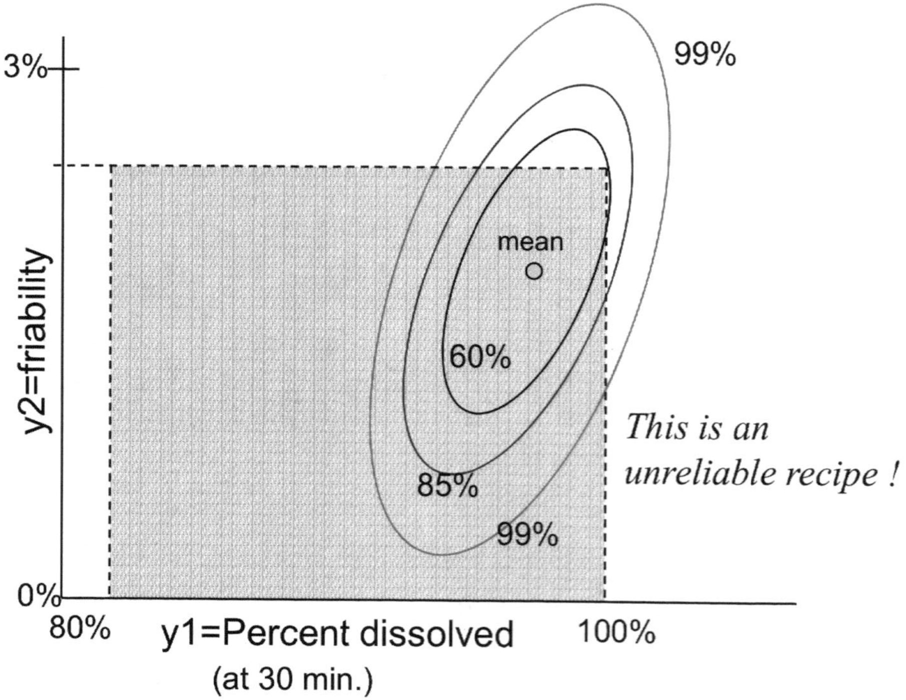


Prob(acceptable result)=0.65

# Design Space

- The Design Space is simply the set of all *reliable* recipes.
- Recall that each recipe is composed of a list of factors such as ingredients or conditions (e.g. mixing speed).
- As such, each recipe can be plotted as a point in space.

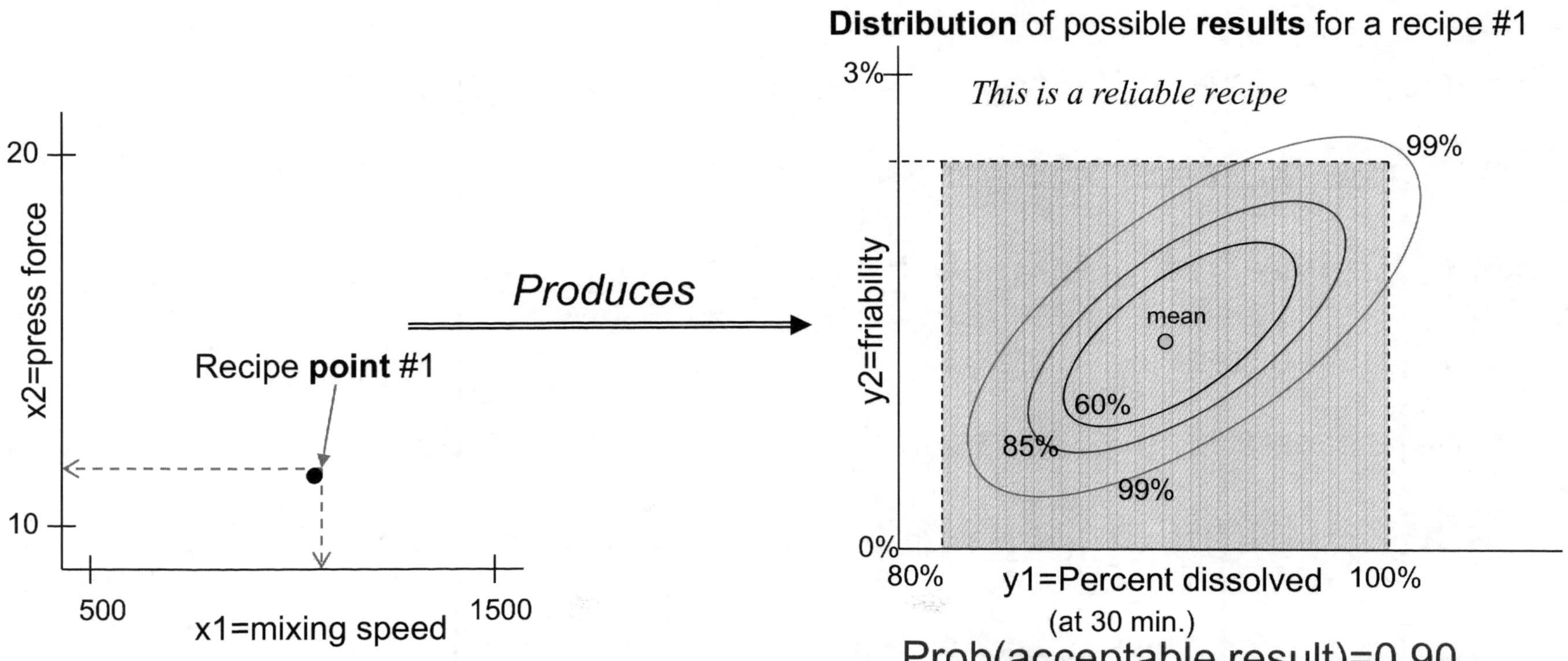


Prob(acceptable result)=0.90

# Design Space

- The Design Space is simply the set of all *reliable* recipes.
- Each recipe point produces a results distribution that is associated with a probability of acceptance.

Recipe point #1 ⇨ Distribution #1 ⇨ Probability #1 of acceptance

Recipe point #2 ⇨ Distribution #2 ⇨ Probability #2 of acceptance

Recipe point #3 ⇨ Distribution #3 ⇨ Probability #3 of acceptance

⋮ ⋮ ⋮

- If we collect all of the recipe points together we can make a "heat map" of the associated probabilities of acceptance.
- A Design Space could be the collection of reliable (recipe) points inside the red oval

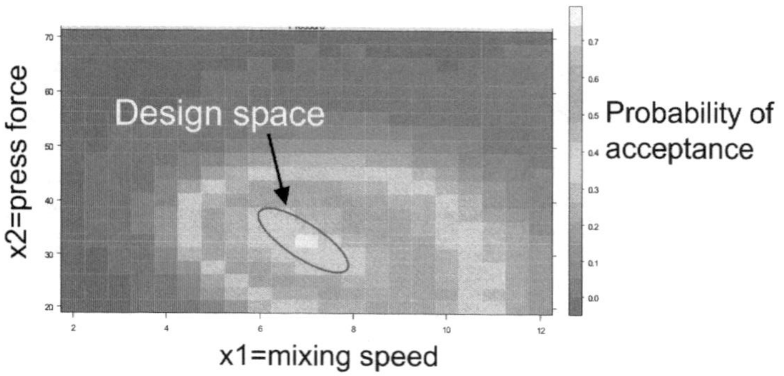

# Design Space

- The Design Space is simply the set of all *reliable* recipes.

- Note that the <u>mean</u> responses for *both* recipes below meet product specifications.

Distribution of possible results for a recipe #1

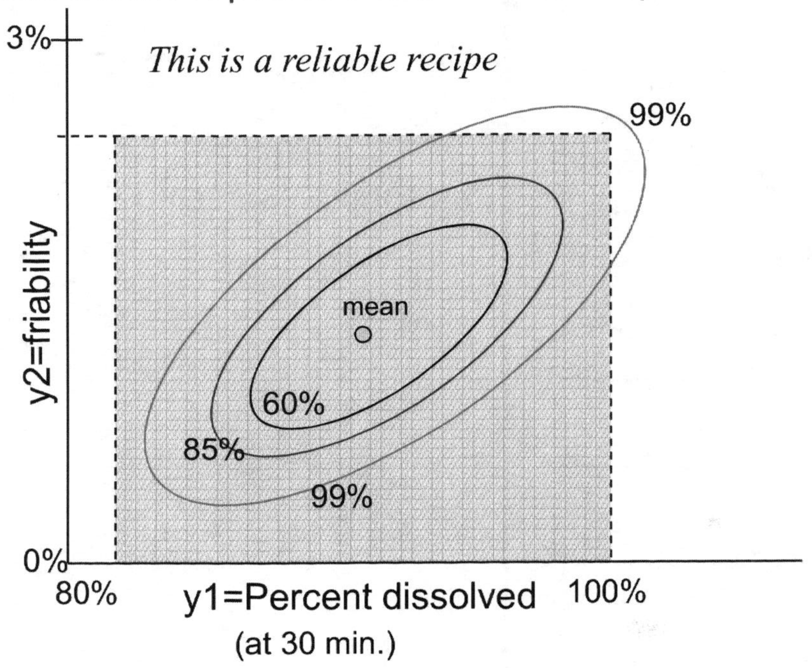


Distribution of possible results for recipe #2

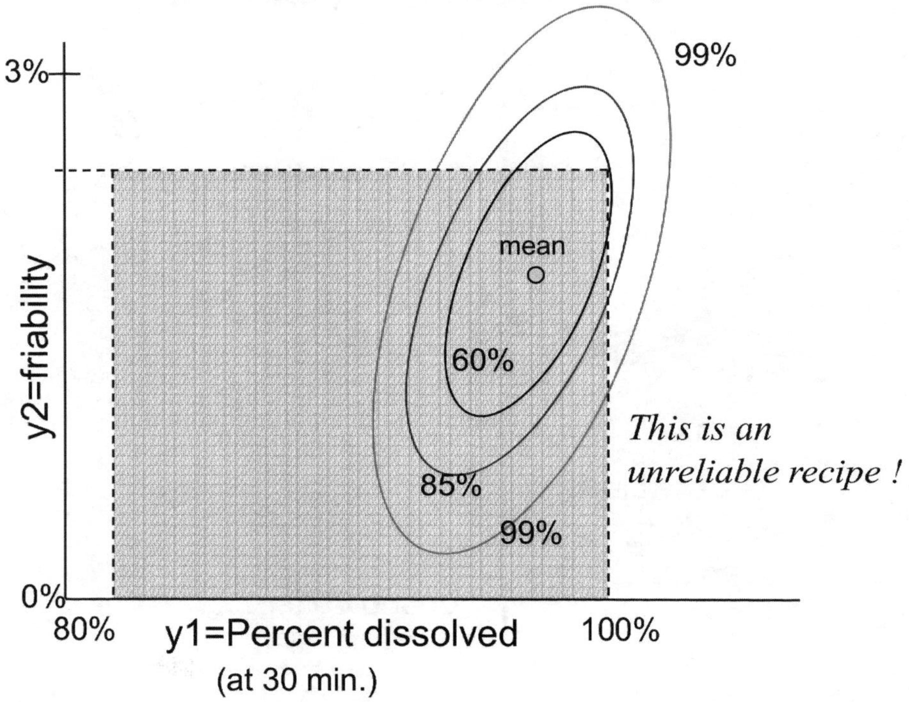

# Why Multivariate Predictive **Distributions** are Needed for a Design Space

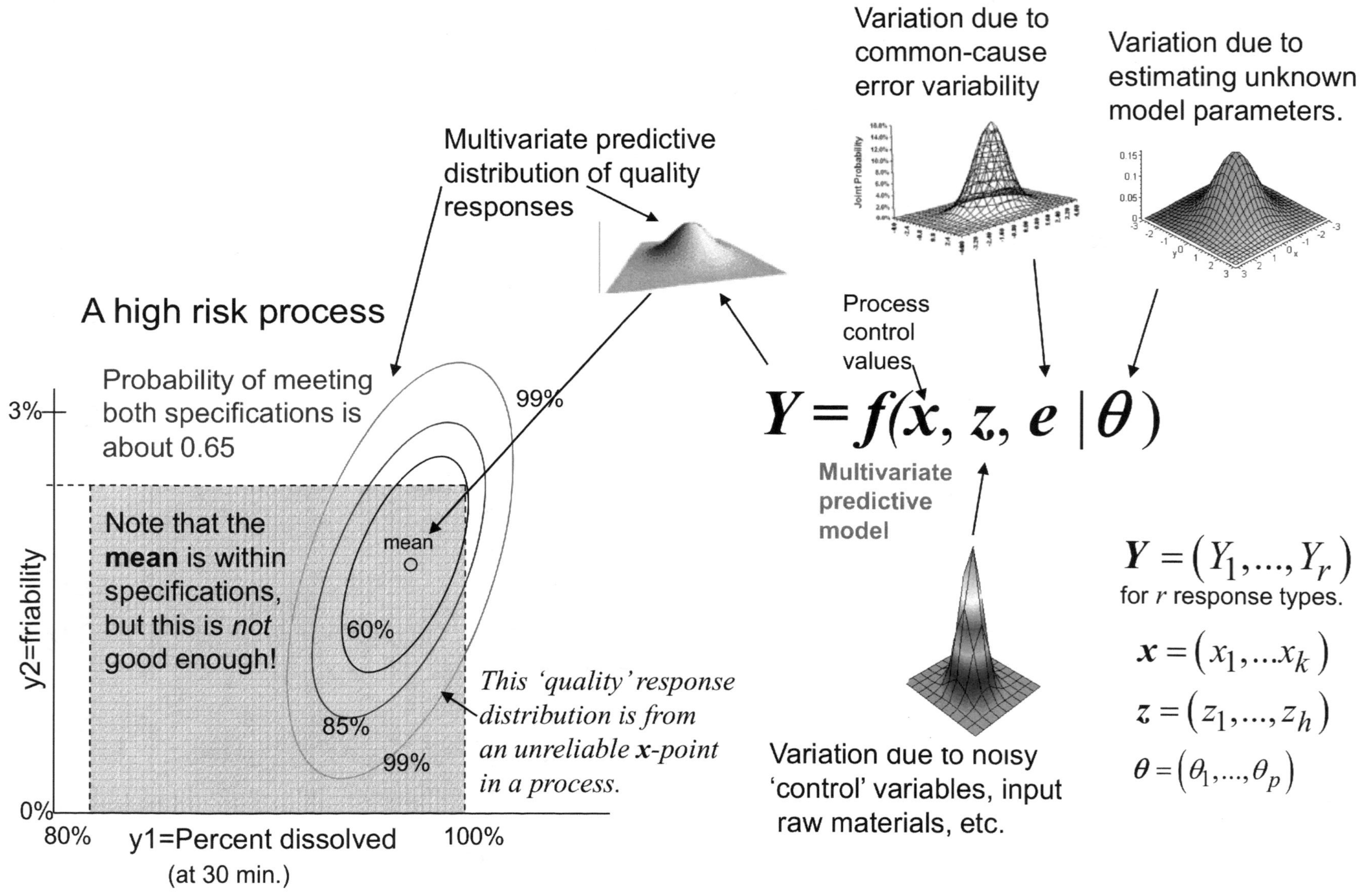

# Why Multivariate Predictive **Distributions** are Needed for a Design Space

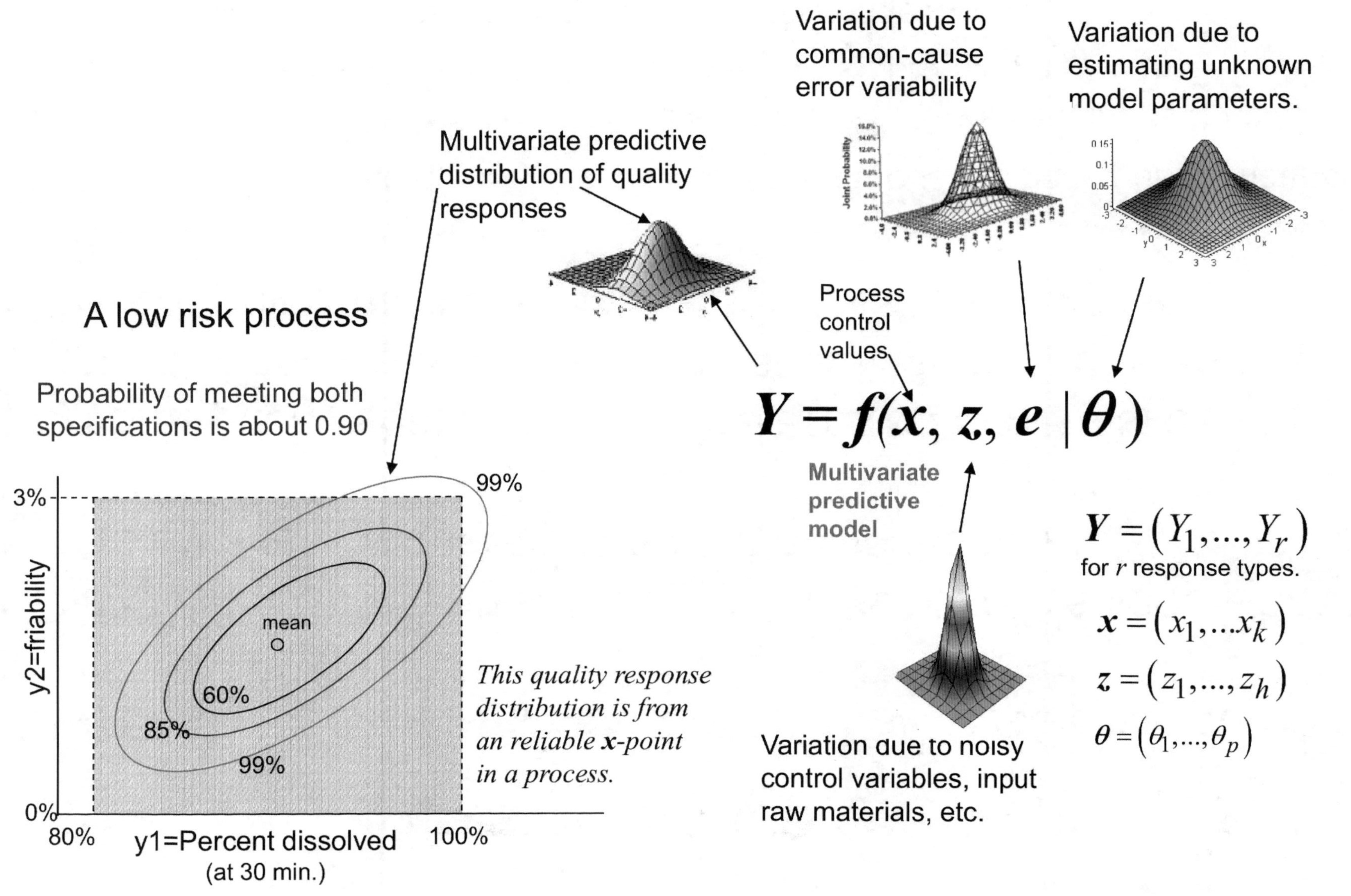

# Design Space

- The Design Space is simply the set of all *reliable* recipes.

- The Design Space should <u>not</u> be the set of all recipes whose results <u>on average</u> meet product specifications.

- Unfortunately, I believe that there are two synergistic factors at work which are steering scientists into looking at "overlapping ***mean*** response surfaces" as a way to construct a Design Space:

  (i) Tradition…commonly used statistical methods are based upon means.

  (ii) Convenience…commercial software for response surfaces provide inferences primarily for means. They do not produce multivariate predictive distributions for the quality responses.

**ICH Q8 Annex:** Design space can be determined from the common region of successful operating ranges for multiple CQA's. The relations of two CQA's, i.e., friability and dissolution, to two parameters are shown in Figures 2a and 2b. Figure 2c shows the overlap of these regions and the maximum ranges of the potential design space.

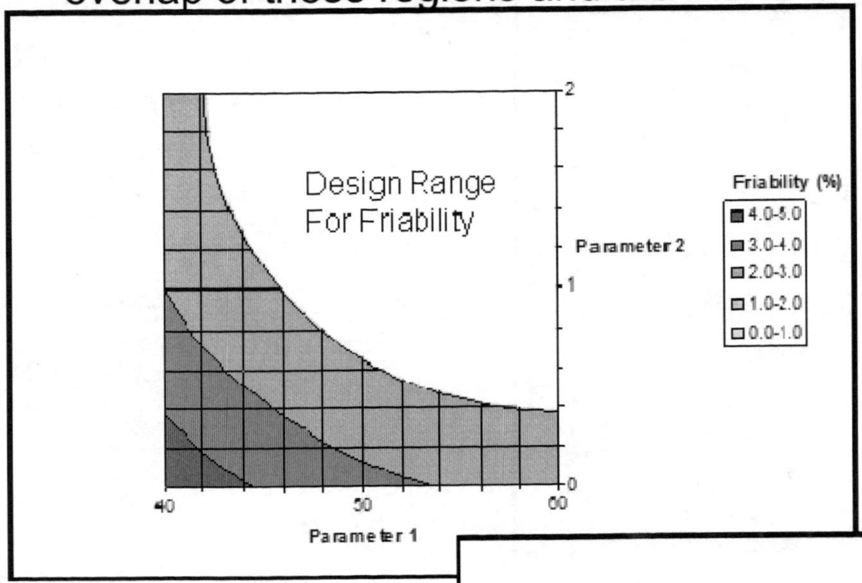


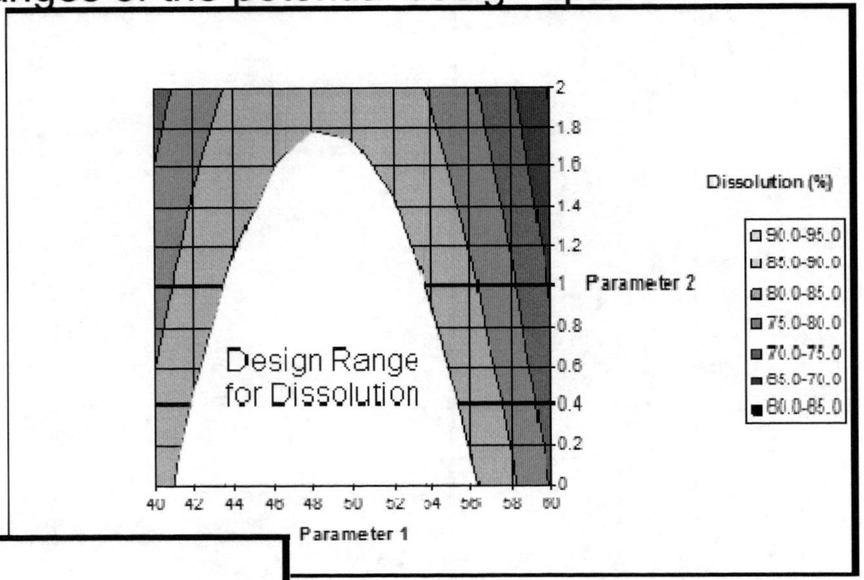


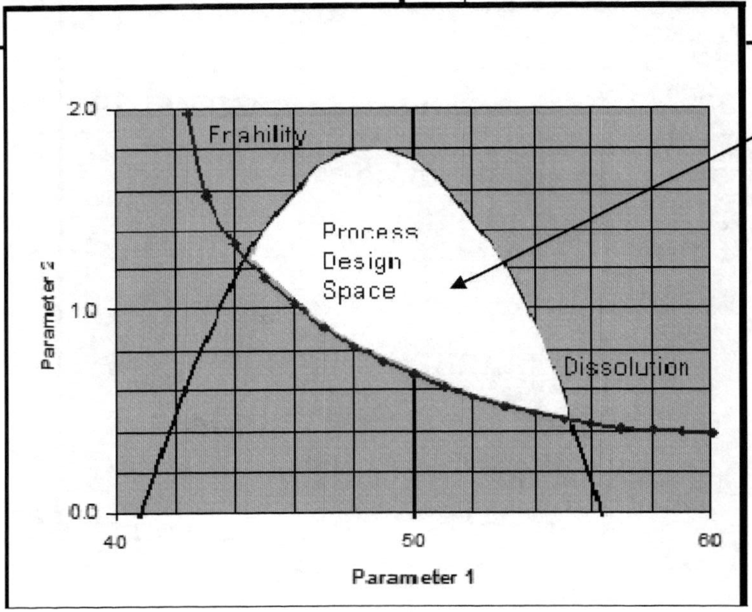


The white region of recipe points corresponds to both responses ***on average*** meeting product specifications.

**This overlapping means surface plot does not quantify "How much assurance?"!**

# Taken from a recent PQLI* Design Space paper

(*J. of Pharm. Innovation*, 2008)

*PQLI = Pharmaceutical Quality Lifecycle Implementation

**This overlapping means surface plot does not quantify "How much assurance?"!**

The yellow region of recipe points corresponds to four responses ***on average*** meeting product specifications.

This plot appears to be from the Design Expert® statistical package.

**Example of Overlay Plot to Identify Impact of Compression Variables on Tablet Properties**

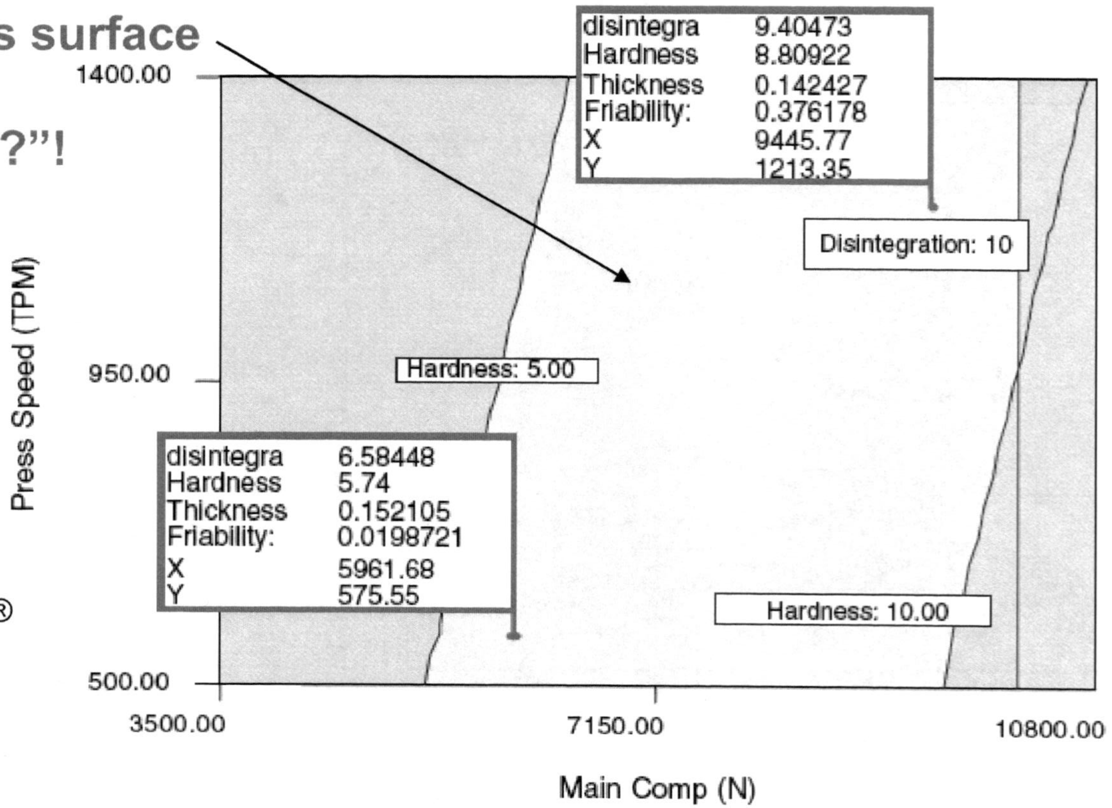

# Design Space Example Due to Greg Stockdale, GSK

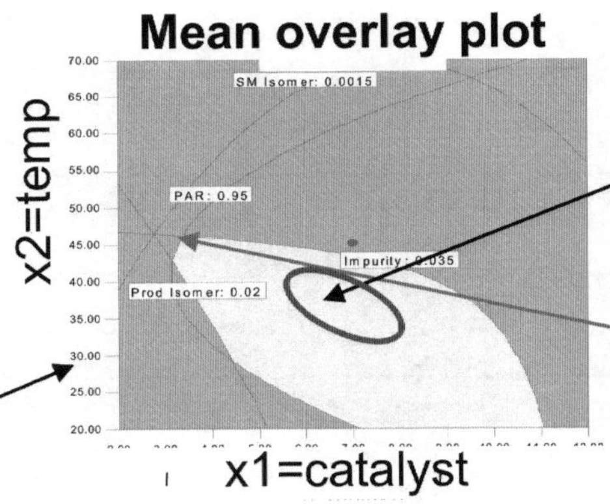


This recipe point has a probability of **0.72** of acceptance.

**But** this recipe point has only a probability of only 0.23 of acceptance!

The yellow region of recipe points corresponds to four responses ***on average*** meeting product specifications.

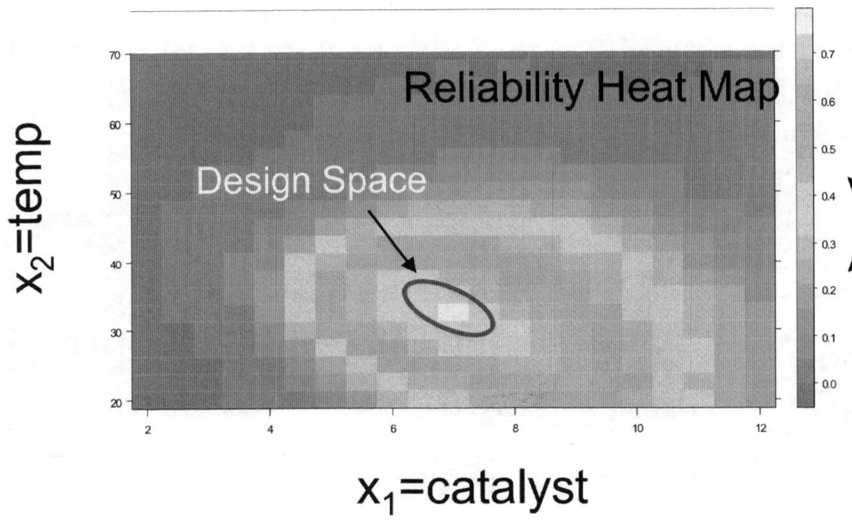


Probability of meeting quality specifications

## The *Best* and *Worst* Posterior Probabilities of Meeting Specifications Found within "Overlapping **Mean** Response" Design Spaces for Six Different Pharmaceutical Experiments with Multiple Quality Responses

| Example Number | Number of factors for the DS | Number of responses for the DS | Minimum posterior probability of acceptance | Maximum posterior probability of acceptance |
|---|---|---|---|---|
| 1 | 4 | 3 | 0.15 | 0.84 |
| 2 | 4 | 4 | 0.23 | 0.72 |
| 3 | 3 | 3 | 0.34 | 1 |
| 4 | 3* | 2 | 0.26 | 0.74 |
| 5 | 4 | 3 | 0.11 | 0.33 |
| 6 | 3 | 4 | 0.11 | 0.84 |

*Mixture experiment. Factors sum to 1.

** Yellow shaded table entries correspond to in-house experiments.
Clear entries for examples from the published pharmaceutical development literature.

# Challenges to Constructing the ICH Q8 Design Space

## Summary

*The challenges….my opinion….*

- Getting researchers to recognize the importance of a probabilistic *model* to quantify "How much assurance?" of meeting product specifications.

- Computational issues. These can be solved with sufficient effort. The Bayesian statistical inference approach provides a unifying paradigm.

- If you see a presentation on Design Space, ask "*How much* assurance is there of meeting product specifications?"

A visual metaphor for a Design Space composed of overlapping mean response surfaces….

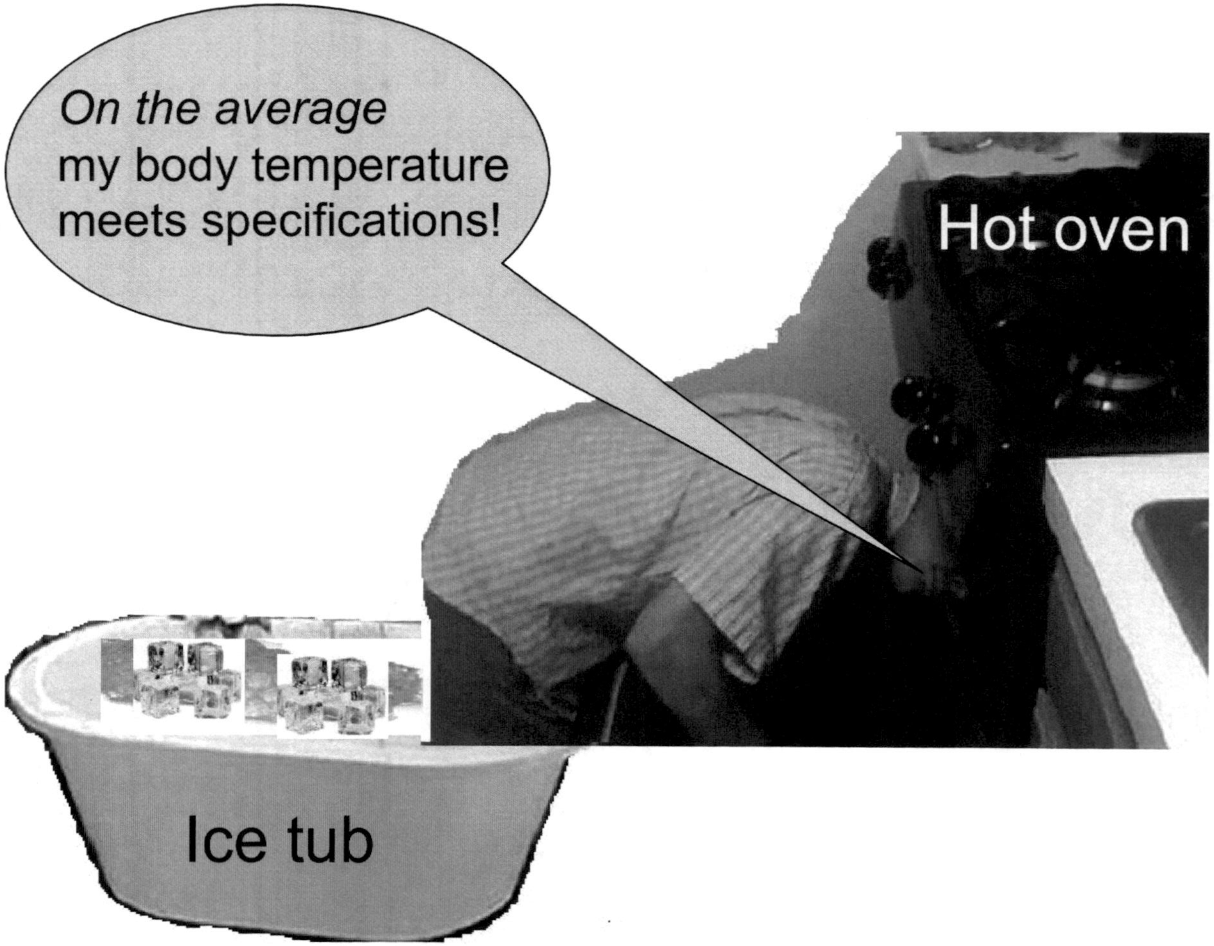

# References


Claycamp, H. G. (2008), "Room for Probability in ICH Q9: Quality Risk Management", presented at the *Pharmaceutical Statistics 2008: Confronting Controversy* conference., March 2008, Arlington, VA. (sponsored by the Institute of Validation Technology).

ICH Q8 (2007), "Pharmaceutical Development Annex to Q8"

Miró-Quesada, G., del Castillo, E., and Peterson, J. J. (2004), "A Bayesian Approach to for Multiple Response Surface Optimization with Noise Variables", *Journal of Applied Statistics*, 31, 251-270.

Peterson, J. J. (2004), "A Posterior Approach to Multiple Response Surface Optimization, *Journal of Quality Technology*,

Peterson, J. J. (2008), "A Bayesian Approach to the ICH Q8 Definition of Design Space", *Journal of Biopharmaceutical Statistics*, 18, 959-975.

Peterson, J. J. Snee, R. D., McAllister, P.R., Schofield, T. L., and Carella, A. J., (2009) "Statistics in the Pharmaceutical Development and Manufacturing" (with discussion), *Journal of Quality Technology*, 41, 111-147.

Peterson, J. J. and Yahyah, M. (2009) "A Bayesian Design Space Approach to Robustness and System Suitability for Pharmaceutical Assays and Other Processes", *Statistics in Biopharmaceutical Research* currently available at http://www.amstat.org/publications/sbr.cfm .

Stockdale, G. and Chen, A. (2009), "Finding Design Space and Reliable Operating Region using a Multivariate Bayesian Approach with Experimental Design", *Quality Technology and Quantitative Management*, currently available at http://www.cc.nctu.edu.tw/~qtqm/ .

# Back-up Slides

# Why Multivariate Predictive **Distributions** are Needed for a Design Space

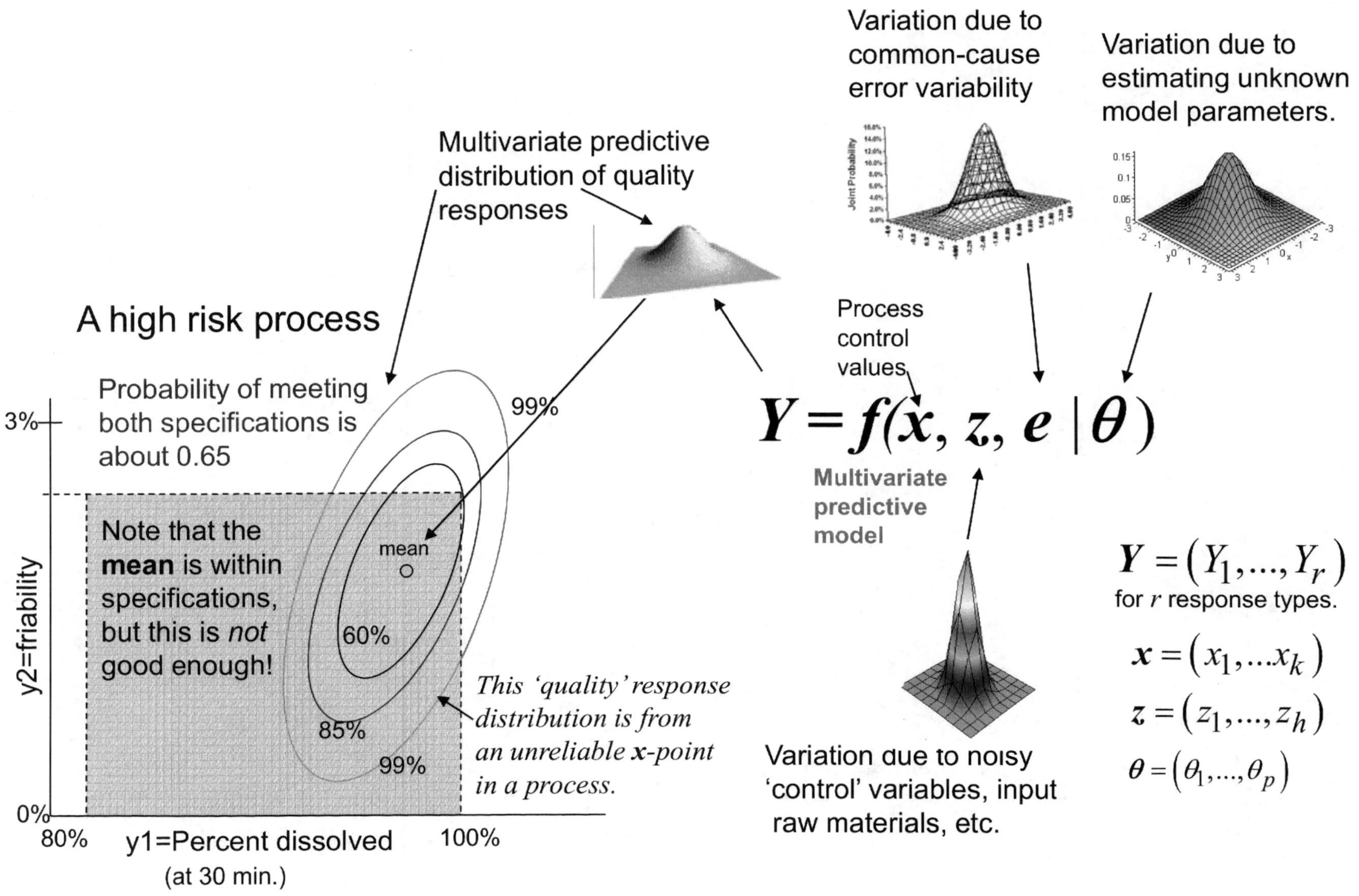

# Why Multivariate Predictive **Distributions** are Needed for a Design Space

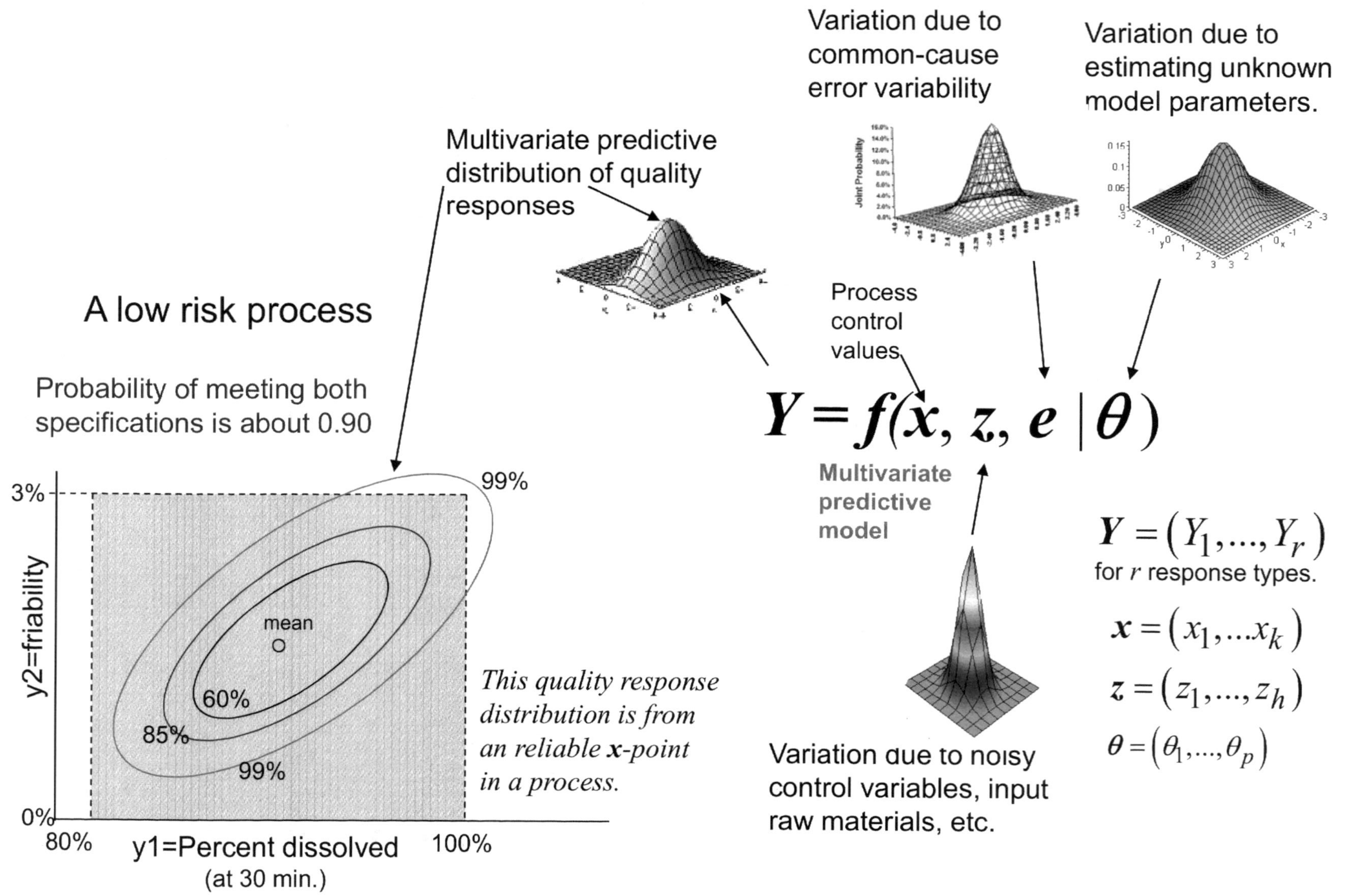

How can we obtain a distribution of unknown model parameters?

There are at least two ways to do this:

- Use Bayesian statistical methodology

- Use “parametric bootstrap” statistical methodology

# The Bayesian approach to obtain a multivariate distribution of unknown model parameters from experimental data

posterior distribution of unknown model parameters

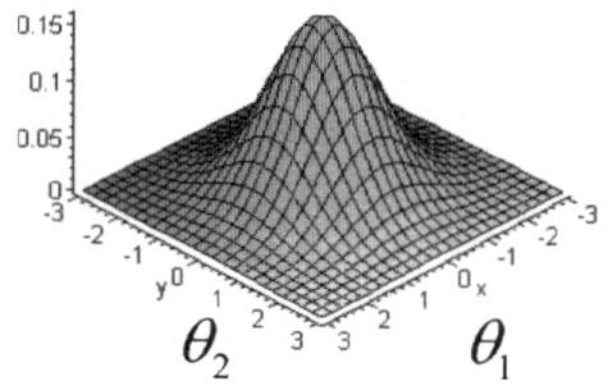


Bayes' rule weighting function

prior distribution of unknown model parameters

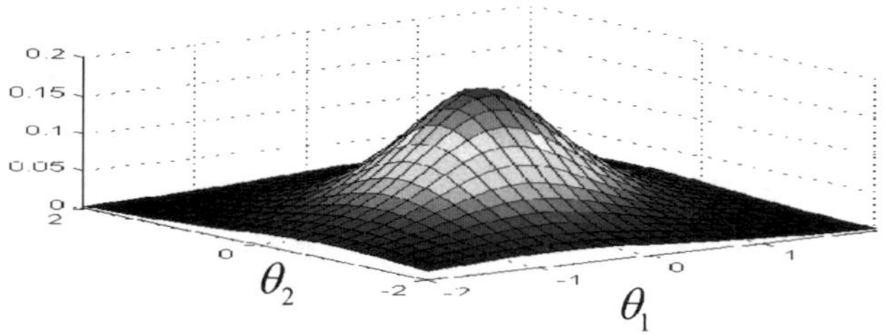


$$p(\boldsymbol{\theta} \mid \text{data}) = \frac{L(data \mid \boldsymbol{\theta})}{\int L(data \mid \boldsymbol{\theta}) \pi(\boldsymbol{\theta}) d\boldsymbol{\theta}} \times \pi(\boldsymbol{\theta})$$

- A probability procedure known as "Markov Chain Monte Carlo" (**MCMC**) can be used to obtain the posterior distribution of unknown model parameters given experimental data (and possibly prior information).

## A "simple" parametric bootstrap approach to obtain a multivariate distribution of unknown model parameters from experimental data

1. Fit model to the data to obtain an estimate, $\hat{\theta}$, of the vector of unknown model parameters, $\theta$.

2. Use $\hat{\theta}$ to simulate many new sets of "data": $\boldsymbol{Y}_1^{(1)}, \ldots, \boldsymbol{Y}_n^{(1)}$ data set 1

   $\vdots$

   $\boldsymbol{Y}_1^{(10,000)}, \ldots, \boldsymbol{Y}_n^{(10,000)}$ data set 10,000

3. Use <u>each</u> new data set to obtain a new parameter estimate: $\hat{\boldsymbol{\theta}}^{(1)}, \ldots, \hat{\boldsymbol{\theta}}^{(10,000)}$

4. The set of parameter estimates in 3. above forms a distribution that reflects the uncertainty of the unknown model parameters.

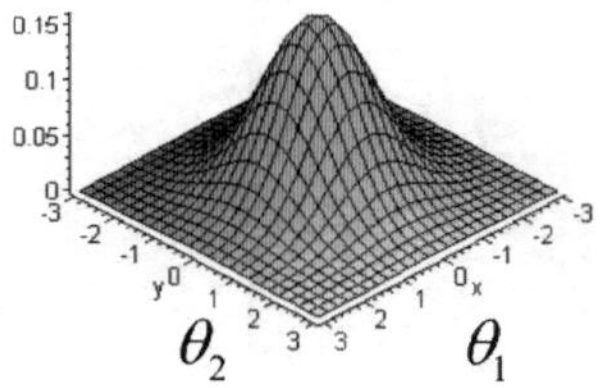

## Overlapping Means vs. Bayesian Reliability Approach to Design Space:

### **An Example** – due to Greg Stockdale, GSK.

*Example:* An intermediate stage of a multi-stage route of manufacture for an Active Pharmaceutical Ingredient (API).

**Measurements:**

- Four controllable quality factors ($x$'s) were used in a designed experiment. (x1='catalyst', x2= 'temperature', x3='pressure', x4='run time'.)

- A (face centered) Central Composite Design (CCD) was employed. (It was a Full Factorial (30 runs), with no aliasing.)

- Four quality-related response variables, $Y$'s, were measured. (These were three side products and purity measure for the final API.)

Y1= 'Starting material Isomer', Y2='Product Isomer', Y3='Impurity #1 Level', Y4='Overall Purity measure'

- Quality Specification limits: Y1<=0.15%, Y2<=2%, Y3<=3.5%, Y4>=95%.

Multidimensional Acceptance region, $A = [0, 0.0015] \times [0, 0.02] \times [0, 0.035] \times [0.95, 1]$

Overlapping Means vs. Bayesian Reliability Approach to Design Space:

**An Example** – due to Greg Stockdale, GSK.

| **Prediction Models:** | Model Terms | | | | | | | | | | | | | | |
|---|---|---|---|---|---|---|---|---|---|---|---|---|---|---|---|
| Response | x1 | x2 | x3 | x4 | x11 | x22 | x33 | x44 | x12 | x13 | x14 | x23 | x24 | x34 | |
| SM Isomer | Δ | Δ | Δ | Δ | | Δ | Δ | Δ | | | | Δ | | | |
| Prod Isomer | Δ | | Δ | Δ | | | Δ | | | Δ | | | | | |
| Impurity | Δ | | Δ | | | | | | | Δ | | | | | |
| Purity | Δ | | Δ | | Δ | | | | | | | | | | |

Temperature = x1 Pressure = x2 Catalyst Amount = x3 Reaction time = x4

# An Overlapping means approach to Design Space for an Active Pharmaceutical Ingredient (API)

## The Design Space is the "Sweet Spot" Highlighted in Yellow below

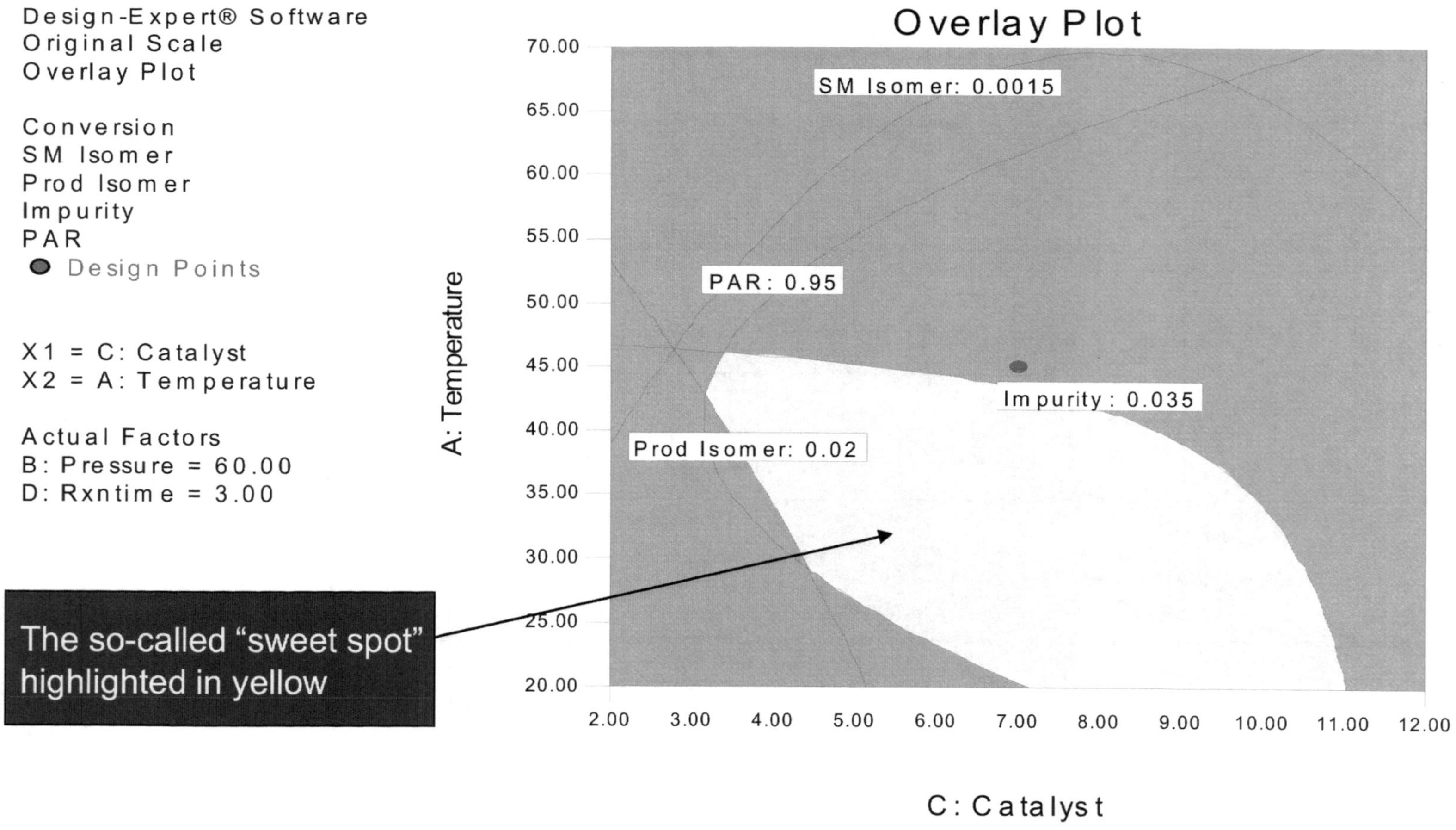

# Why is the "sweet spot" not so sweet?

- If the mean of $Y$ at a point $\boldsymbol{x}$ is less than an upper bound, $u$,

then all that guarantees is that $Pr(Y \le u \mid \boldsymbol{x}) > 0.5$

(For $\Pr(l \le Y \le u \mid \boldsymbol{x})$ there is no guarantee that $\Pr(l \le Y \le u \mid \boldsymbol{x}) > 0.5$)

- Suppose $\mu_{Y_1} \le u_1$ and $\mu_{Y_2} \le u_2$ If $Y_1$ and $Y_2$ were independent,

then all that is guaranteed is that $Pr(Y_1 \le u_1, Y_2 \le u_2 \mid \boldsymbol{x}) > 0.25$

For $k$ independent $Y_i$'s the situation becomes:

$$Pr(Y_1 \le u_1, \ldots, Y_k \le u_k \mid x) > 0.5^k$$

- If $Y_1$ and $Y_2$ are positively correlated then it may be easier to find

$\boldsymbol{x}$-points to make $Pr(Y_1 \le u_1, Y_2 \le u_2 \mid \boldsymbol{x})$ large. Likewise, if $Y_1$ and $Y_2$

are negatively correlated (for each $\boldsymbol{x}$) then it may be more difficult.

Note: Corr($Y_3$, $Y_4 \mid \boldsymbol{x}$) is about -0.8 for the API experiment.

# Why is a multivariate reliability approach needed?

(Accounting for correlation among the responses…a simple example)

- Suppose we have a process with four key responses, $Y_1$, $Y_2$, $Y_3$, $Y_4$
- For simplicity, let's assume that $Y = (Y_1, Y_2, Y_3, Y_4)' \sim N(\mathbf{0}, \boldsymbol{\Sigma})$
- Let $A = (-\infty, 1] \times (-\infty, 1] \times (-\infty, 1] \times (-\infty, 1]$
- Consider $\Pr(\boldsymbol{Y} \in A)$ If $\boldsymbol{\Sigma} = \boldsymbol{I}$, then $\Pr(\boldsymbol{Y} \in A) = 0.50$

But if

$$\boldsymbol{\Sigma} = \begin{bmatrix} 1 & 0.9 & 0.9 & 0.9 \\ 0.9 & 1 & 0.9 & 0.9 \\ 0.9 & 0.9 & 1 & 0.9 \\ 0.9 & 0.9 & 0.9 & 1 \end{bmatrix}$$

then $\Pr(\boldsymbol{Y} \in A) = 0.75$

and if

$$\boldsymbol{\Sigma} = \begin{bmatrix} 1 & -0.318812 & -0.289382 & -0.199902 \\ -0.318812 & 1 & -0.392855 & -0.356863 \\ -0.289382 & -0.392855 & 1 & -0.29417 \\ -0.199902 & -0.356863 & -0.29417 & 1 \end{bmatrix}$$

then $\Pr(Y \in A) = 0.42$

## Overlapping Means vs. Bayesian Reliability Approach to Design Space: **An Example**

*Overlapping Mean Response Surface Approach* –

- Can be computed using SAS/JMP, Design Expert, Minitab, etc.

- The "sweet spot" region is determined by the overlapping mean response surfaces that are all simultaneously within their specification limits.

- However the overlapping mean response approach:
  (i) Does not take into account the model parameter uncertainty
  (ii) Does not provide a measure of assurance to say "How likely it is that future responses will meet their specifications."
  (iii) Does not take into account the correlation structure of the multivariate distribution of future responses.

- Thus the overlapping means approach does not address the question begged by the ICH Q8 definition of Design Space…namely, "How much assurance do we have of meeting process quality specifications?"

## Overlapping Means vs. Bayesian Reliability Approach to Design Space: **An Example**

*A Bayesian Reliability Response Surface Approach* –

- A posterior predictive approach:
  (i) Takes into account the model parameter uncertainty
  (ii) Provides a measure of assurance to say "How likely it is that future responses will meet their specifications."
  (iii) Takes into account the correlation structure of the multivariate distribution of future responses.

- Thus the Posterior predictive approach addresses the question begged by the ICH Q8 definition of Design Space…namely, "How much assurance do we have of meeting process quality specifications?"

- Both graphical and tabular approaches can be used with the posterior predictive approach to better understand the resulting Design Space.

# Design Space Table of Computed Reliabilities[1] for the API (sorted by joint probability[2])

Note that the largest probability of meeting specifications is only about 0.75

| Temp | Pressure | Catalyst | Rxntime | Joint Prob | SM Isomer Prob | Prod Isomer Prob | Impurity Prob | Purity Prob |
|---|---|---|---|---|---|---|---|---|
| 35 | 60 | 6 | 3 | 0.752 | 1 | 0.9985 | 0.8435 | 0.79 |
| 32.5 | 60 | 7 | 3 | 0.743 | 1 | 0.9995 | 0.7875 | 0.8295 |
| [illegible] | [illegible] | [illegible] | 3 | 0.7375 | 0.9995 | 0.9995 | 0.7855 | 0.8255 |
| [illegible] | [illegible] | [illegible] | 3 | 0.737 | 1 | 0.9975 | 0.821 | 0.7845 |
| [illegible] | [illegible] | [illegible] | 3 | 0.7335 | 1 | 0.9995 | 0.7775 | 0.8175 |
| 37.5 | 60 | 6.5 | 3 | 0.725 | 1 | 1 | 0.7485 | 0.845 |
| 35 | 60 | 6.5 | 3 | 0.7225 | 1 | 1 | 0.77 | 0.812 |
| 32.5 | 60 | 6 | 3 | 0.7195 | 1 | 0.9955 | 0.864 | 0.7415 |
| 30 | 60 | 7 | 3 | 0.717 | 1 | 0.999 | 0.8075 | 0.759 |
| 32.5 | 60 | 7.5 | 3 | 0.716 | 1 | 1 | 0.734 | 0.859 |
| 37.5 | 60 | 5.5 | 3 | 0.7145 | 1 | 0.993 | 0.8065 | 0.7565 |
| 35 | 60 | 7 | 3 | 0.712 | 1 | 1 | 0.731 | 0.8555 |

**Optimal Reaction Conditions**

Marginal Probabilities

[1] This is only a small portion of the Monte Carlo output.

[2] values were computed using SAS IML

# Summary

1. Some point & click DoE software applications make it easy to construct a region of overlapping mean responses.

2. However, such overlapping regions of ***means*** may produce a design space that is much too large, harboring operating conditions with a low probability of meeting all of the specification limits.

3. What is needed is a reliability-based "response" surface function of the form

$$p(\boldsymbol{x}) = \mathrm{Prob}(\boldsymbol{Y} \text{ is in } A \mid \text{given } \boldsymbol{x} \text{ and the data}),$$

where $\boldsymbol{Y}=(Y_1,\ldots,Y_r)$ are the quality responses and $A$ is the acceptance region formed by the process specifications.

4. The Bayesian Design Space is just the set of all controllable process conditions (i.e. $\boldsymbol{x}$-points) such that $p(\boldsymbol{x})$ is at least some prespecified reliability level.

# Correlating Powder Flowability to Particle Size Distribution Using Chemometric Techniques

Christopher L. Burcham, Chemical Product Research and Development; Process Definition, Eli Lilly and Company, Indianapolis, IN

Robert T. Roginski, Eigenvector Research, Inc., Wenatchee, WA

A compound currently in Phase III clinical trials is being developed as a 140 mg dose. In order to meet the needs of the patient population, the tablet must be no larger than 450 mg, resulting in a high drug substance loading. The drug product is manufactured using a roller compaction process. The high drug load requires that the drug substance have good flow properties as a dry powder. An understanding of the impact that the particle size distribution of the drug substance on the powder flowability is needed to allow for the establishment of particle size specifications on the drug substance. Principal Components Analysis PCA) was applied to the particle size distribution data generated from numerous crystallization conditions and scales to identify clustering. A predictive model for powder flowability was built based on the particle size distribution data using Partial Least Squares (PLS). This model was further refined using local weighting techniques to deliver better prediction metrics. The results of these modeling efforts are presented.

See more of this Session: 11 Empirical Modeling and Experimental QbD
See more of this Group/Topical: Topical I: Comprehensive Quality by Design in Pharmaceutical Development and Manufacture

# Predictive Modeling of Pharmaceutical Processes with Missing and Noisy Data

**Fani Boukouvala**, Chemical and Biochemical Engineering, Rutgers University, Piscataway, NJ

Fernando Muzzio, Chemical and Biochemical Engineering, Rutgers University, Piscataway, NJ

M.G. Ierapetritou, Chemical & Biochemical Engineering Department, Rutgers University, the State University of New Jersey, Piscataway, NJ

Current trends in pharmaceutical product development focus on the fundamental understanding and characterization of all process unit operations, which will lead to the development of predictive models in order to minimize the variability in product performance. The pharmaceutical industry is one of the most characteristic examples where high quality products- with strict predefined performance specifications - must be produced. The successful completion of this task is obstructed by the fact that mechanical and physiochemical properties of pharmaceutical APIs and excipients used are not completely understood [1]. Other industrial fields (food, ceramics, catalyst manufacturing etc)- that deal with processing of materials in powder form- are faced with the same challenging issue simply because powders share both characteristics of fluids and solids. It is very difficult to model the behavior of granular materials using first principles as they undergo different key processes, such as feeding, mixing and tablet pressing, without making a series of assumptions and introducing a relative amount of uncertainty. Data driven modeling methods on the other hand, connect the system state variables (input, internal and output variables) and use only a limited knowledge of the physical behavior of the system and the available experimental data. Data analysis techniques however are required to "prepare" the data sets used for modeling in order to minimize uncertainty. Data analysis poses a challenging set of problems such as: high-dimensionality, computationally expensive objectives and incomplete or missing information.

Until now, several models have been proposed for individual process operations. Methods such as Monte Carlo simulations, particle-dynamic simulations, heuristic models, and kinetic theory models have been proposed to characterize powder flow during the process of mixing using continuous blenders. Based on the continuum approximation, Sudah et al. developed a dispersion model to describe flow of powders in continuous blenders [2]. Berthiaux et al. proposed a model of continuous mixing based on the theory of Markov chains [3], while Ghaderi investigated axial mixing in continuous mixers and proposed a model using the Danckwerts- Weinekotter formula [4]. A number of papers have employed the discrete-element method (DEM) to model the behavior of granular flow in continuous mixers. The drawback to these simulations, however, is their computational efficiency. In the work of Portillo et al. it is observed that as the number of particles in the system increase, the computational intensity of a DEM simulation becomes prohibitive [5]. Work has also been done to improve the design of feeders ([6],[7]), however, limited attention has been directed at the analytical study of feeder performance in terms of the input process variables and operating conditions. Unlike all the above examples of first principle based models, data driven methods were used by Jia et al. to develop a less computationally expensive model based on experimental data alone [8]. Specifically, Kriging and Response Surface methodologies were used to predict the output flow variability of a loss-in-weight feeder and the residence time distribution of a continuous mixer. It is however known that the uniformity of the output flow of a feeding unit affects the behavior and output of the blender and consequently the properties of the final product (tablet). No work has yet been done in terms of designing an integrated model that will relate both properties of the material processed and the operating parameters to the properties of the final product.

Design space in pharmaceutical development is defined as collection of all multidimensional combinations of input variables, which characterize material attributes (composition, cohesion, size distribution etc) and processing parameters (flow rate, mixer fill level, agitator speed, blade pattern, residence time etc). Thus, large multivariate data matrices are our resources for creating models, which will capture the effects of multiple variables to the final desired output- as well as interactions and correlations between variables.

Working towards combining all the collected data, in order to produce an integrated model, we are often faced with the problem of having to work with incomplete data sets. Missing information may often arise from experiments due to insufficient sampling, errors during data acquisition or even when data fusion techniques are required in order to combine information from different sources. By simply discarding the missing observations, valuable information for subsequent analysis of the data can be lost. An exploratory analysis of the popular imputation algorithms that exist in literature is done by Huang et. al in order to compare their performances and suggest guidelines for the selection of the appropriate algorithm for a specific application [9].

In this work, different imputation techniques such as mean imputation, k-nearest neighbor imputation and multiple imputation are incorporated to the predictive model algorithm of pharmaceutical unit operations. Experimental data is obtained for different unit processes of tablet manufacturing (feeding, mixing and tablet press) and used for a thorough assessment of the efficiency and effect of the different methodologies on the average error of prediction of output variables. It is observed that different methods can be appropriate for different percentages of missing information. A more sophisticated model based method is formulated based on Maximum Likelihood Estimation (Expectation- Maximization algorithm - EM) to estimate parameters of the incomplete data sets and thus "fill out" the missing values based on the position they would "most likely" be found. Principal Component Analysis is combined with the EM algorithm, in order to produce a hybrid method of imputing missing values. This approach was first introduced by Andrews and Wentzell who demonstrated that principal components extracted with up to 10% of missing information were comparable to the projections of the complete data [10]. This iterative procedure is proposed because it can overcome the following barriers:

·Reduction of dimensionality by sorting the variables based on the correlations (covariance) between them,

·Minimization of noise,

·Imputation of missing values.

Cross-validation and bootstrap techniques are used for validation of model selection and further accuracy estimation [11].

Data- driven predictive models have already shown very promising results in characterizing the behavior of separate unit operations of pharmaceutical product development [8]. Our ultimate goal is to gradually increase the complexity of the predictive model by allowing it to have as many input and output variables as necessary, in order to design the optimal set of process parameters that will result in the optimal performance given a set of material properties. Such a model will ideally ensure product quality and manufacturing cost reduction without having computationally restricting requirements. However, the key to successful modeling using such techniques is the right preparation of the data, which includes recognition of outliers and imputation of missing values. The proposed algorithm will be able to identify the appropriate imputation method that will minimize the uncertainty of missing and unreliable information- depending on the sequence and amount of missingness. Incorporating these methods into interpolatory regression based models- such as Kriging and Response Surface methods- will be a very important step towards the completion of the desired integrated model.

References:

1. Portillo, P., et al., Quality by Design Methodology for Development and Scale-up of Batch Mixing Processes. Journal of Pharmaceutical Innovation, 2008. 3(4): p. 258-270.

2. Sudah, O.S., et al., Quantitative characterization of mixing processes in rotary calciners. Powder Technology, 2002. 126(2): p. 166-173.

3. Berthiaux, H., Marikh, K., Mizonov, V., Ponomarev, D., Barantzeva, Modeling continuous powder mixing by means of the theory of Markov chains. Particulate Science and Technology 2004. 22: p. 379-389.

4. Ghaderi, A., On characterization of continuous mixing of particulate materials. Particulate Science and

Technology, 2003. 21: p. 271-282.

5. Patricia M. Portillo, F.J.M., Marianthi G. Ierapetritou,, Hybrid DEM-compartment modeling approach for granular mixing. AIChE Journal, 2007. 53(1): p. 119-128.

6. Gundogdu, M.Y., Design improvements on rotary valve particle feeders used for obtaining suspended airflows. Powder Technology, 2004. 139(1): p. 76-80.

7. Reist, P.C. and L. Taylor, Development and operation of an improved turntable dust feeder. Powder Technology, 2000. 107(1-2): p. 36-42.

8. Jia, Z., Davis,E., Muzzio, F., Ierapetritou, M.,, Predictive Modeling for Mixing and Feeding Processes using Kriging and Response Surface. In Press.

9. Huang, S.H., R. Kothamasu, and N. Rapur, Iterative imputation algorithms for process modeling with incomplete data. Intelligent Data Analysis, 2007. 11(2): p. 189-202.

10. Andrews, D.T. and P.D. Wentzell, Applications of maximum likelihood principal component analysis: incomplete data sets and calibration transfer. Analytica Chimica Acta, 1997. 350(3): p. 341-352.

11. Kohavi, R. A Study of Cross-Validation and Bootstrap for Accuracy Estimation and Model Selection. in IJCAI. 1995.

See more of this Session: 11 Empirical Modeling and Experimental QbD
See more of this Group/Topical: Topical I: Comprehensive Quality by Design in Pharmaceutical Development and Manufacture

# Nonlinear Programming Based Methods for PCA Parameter Estimation Using Data with Missing Elements

Rodrigo López-Negrete de la Fuente, Chemical Engineering, Carnegie Mellon University, Pittsburgh, PA

Salvador Garcia-Munoz, Process Modeling & Engineering Technology, Pfizer Global Research & Development, Groton, CT

Lorenz T. Biegler, Chemical Engineering, Carnegie Mellon University, Pittsburgh, PA

Abstract

Many chemical and pharmaceutical products are produced in plants where there is a large amount of sensors to extract data for analysis. In such cases, where large amounts of possibly correlated data is available for analysis, Principal Component Analysis (PCA) as well as other multivariate methods are commonly used. PCA is a multivariate technique in which a large number of related variables is transformed into a possibly smaller number of uncorrelated variables.[1]

Unfortunately, there can be cases in which there are missing values in the data used for analysis. This could be due to failing sensors, experiments used to obtain the data could be prohibitely expensive, or in the case when historical data is being used it may get lost.[2] In these situations different methods for obtaining the model parameters can be used, mainly a modified version of the Nonlinear Iterative Partial Least Squares (NIPALS) algorithm or the Expectation Maximization (EM) algorithm.[3] The modified NIPALS is the preferred method of the two because it is faster. However, there are some caveats associated with its use: the obtained loadings are not orthogonal and the variance of the score vectors might be greater than the variance of the original data (after mean centering and auto scaling). To address this issues the EM algorithm introduces the use of imputed values for the missing data. Thus, it solves full data problems using NIPALS at each iteration, then the imputed values are updated with the predictions, and this is repeated until the change in the imputed values is less than a user defined tolerance.[4] This method works well, but it is not very popular because it takes a long time to converge. In the open literature work has been done to analyze the effects of the amounts of missing data and its structure on the PCA models built using different methods.[5,6] Also, once a model has been constructed it can be used for prediction and monitoring. Several papers dealing with prediction with incomplete measurements using PCA and Partial Least Squares (PLS) models have also been written (e.g., Nelson et al. [7], Arteaga and Ferrer [8], and Nelson et al. [9]).

In this work we propose a Nonlinear Programming (NLP) based method for obtaining the parameters for the PCA model in the presence of missing values. By solving a constrained NLP the problems associated with the modified NIPALS algorithm are eliminated. Moreover, this method solves directly the minimization of the squared error in the prediction, and thus, it eliminates the necessity to solve multiple PCA problems with different imputed values. This reduces the amount of calculations required to obtain the parameters as in the EM algorithm, but we still obtain the optimal solution in the least squares sense. This approach is illustrated using randomly generated data, as well as, with data obtained from the pharmaceutical industry.

References

[1] Wold, H. Estimation of Principal Components and Related Models by Iterative Least Squares. In Krishnaiah, P., ed., Multivariate Analysis. Academic Press, New York, 1966, pp. 391-420.

[2] Little, R. J. A.; Rubin, D. B. Statistical Analysis with Missing Data. Second ed. Wiley-Interscience, 2002.

[3] Grung, B.; Manne, R. Missing Values in Principal Component Analysis. Chemom. Intell. Lab. Syst. 1998, 42, 125.

[4] Adams, E.; Walczak, B.; Vervaet, C.; Risha, P. G.; Massart, D. Principal Component Analysis of Dissolution Data with Missing Elements. Int. J. Pharm. 2002, 234, 169.

[5] Walczak, B.; Massart, D. L. Dealing with Missing Data: Part I. Chemom. Intell. Lab. Syst. 2001, 58, 15.

[6] Walczak, B.; Massart, D. L. Dealing with Missing Data: Part II. Chemom. Intell. Lab. Syst. 2001, 58, 29.

[7] Nelson, P. R. C.; Taylor, P. A.; MacGregor, J. F. Missing Data Methods in PCA and PLS: Score Calculations with Incomplete Observations. Chemom. Intell. Lab. Syst. 1996, 35 (1), 45.

[8] Arteaga, F.; Ferrer, A. Dealing with Missing Data in MSPC: Several Methods, Different Interpretations, Some Examples. J. Chemom. 2002, 16, 408.

[9] Nelson, P. R. C.; MacGregor, J. F.; Taylor, P. A. The Impact of Missing Measurements on PCA and PLS Prediction and Monitoring Applicatons. Chemom. Intell. Lab. Syst. 2006, 80 (1), 1.

See more of this Session: 11 Empirical Modeling and Experimental QbD
See more of this Group/Topical: Topical I: Comprehensive Quality by Design in Pharmaceutical Development and Manufacture

# AUTHOR INDEX

# AUTHOR INDEX